ENERGY SCIENCE, ENGINEERING AND TECHNOLOGY

ELECTRICITY RELIABILITY IN THE UNITED STATES

SELECT RESEARCH

ENERGY SCIENCE, ENGINEERING AND TECHNOLOGY

Additional books in this series can be found on Nova's website under the Series tab.

Additional e-books in this series can be found on Nova's website under the e-book tab.

ENERGY POLICIES, POLITICS AND PRICES

Additional books in this series can be found on Nova's website under the Series tab.

Additional e-books in this series can be found on Nova's website under the e-book tab.

ENERGY SCIENCE, ENGINEERING AND TECHNOLOGY

ELECTRICITY RELIABILITY IN THE UNITED STATES

SELECT RESEARCH

EMMANUEL MARTINEZ
EDITOR

nova publishers
New York

For permission to use material from this book please contact us:
Telephone 631-231-7269; Fax 631-231-8175
Web Site: http://www.novapublishers.com

NOTICE TO THE READER

Additional color graphics may be available in the e-book version of this book.

Library of Congress Cataloging-in-Publication Data

ISBN: 978-1-62417-149-9

Published by Nova Science Publishers, Inc. † New York

CONTENTS

PREFACE

Since the 1960s, the U.S. electric power system has experienced a major blackout about once every 10 years. Each has been a vivid reminder of the importance society places on the continuous availability of electricity and has led to calls for changes to enhance reliability. At the root of these calls are judgments about what reliability is worth and how much should be paid to ensure it. In principle, comprehensive information on the actual reliability of the electric power system and on how proposed changes would affect reliability ought to help inform these judgments. Yet, comprehensive, national-scale information on the reliability of the U.S. electric power system is lacking. This book helps to address this gap by assessing trends in U.S. electricity reliability based on information reported by electric utilities on power interruptions experienced by their customers.

Chapter 1 – Since the 1960s, the U.S. electric power system has experienced a major blackout about once every 10 years. Each has been a vivid reminder of the importance society places on the continuous availability of electricity and has led to calls for changes to enhance reliability. At the root of these calls are judgments about what reliability is worth and how much should be paid to ensure it.

In principle, comprehensive information on the actual reliability of the electric power system and on how proposed changes would affect reliability ought to help inform these judgments. Yet, comprehensive, national-scale information on the reliability of the U.S. electric power system is lacking.

This chapter helps to address this information gap by assessing trends in U.S. electricity reliability based on information reported by electric utilities on power interruptions experienced by their customers. The authors' research augments prior investigations, which focused only on power interruptions originating in the bulk power system, by considering interruptions originating both from the bulk power system and from within local distribution systems.The authors' research also accounts for differences among utility reliability reporting practices by employing statistical techniques that remove the influence of these differences on the trends that the authors identify.

The research analyzes up to 10 years of electricity reliability information collected from 155 U.S. electric utilities, which together account for roughly 50% of total U.S. electricity sales. The questions analyzed include:

1. Are there trends in reported electricity reliability over time?
2. How are trends in reported electricity reliability affected by the installation or upgrade of an automated outage management system?
3. How are trends in reported electricity reliability affected by the use of IEEE Standard 1366-2003?

Chapter 2 – Bulk power system reliability is of critical importance to the electricity sector. Complete and accurate information on events affecting the bulk power system is essential for assessing trends and efforts to maintain or improve reliability. Yet, current sources of this information were not designed with these uses in mind. They were designed, instead, to support real-time emergency notification to industry and government first-responders. This chapter reviews information currently collected by both industry and government sources for this purpose and assesses factors that might affect their usefulness in supporting the academic literature that has relied upon them to draw conclusions about the reliability of the US electric power system.

Chapter 3 – Large blackouts, such as the August 14-15, 2003 blackout in the northeastern United States and Canada, focus attention on the importance of reliable electric service. As public and private efforts are undertaken to improve reliability and prevent power interruptions, it is appropriate to assess their effectiveness. Measures of reliability, such as the frequency and duration of power interruptions, have been reported by electric utilities to state public utility commissions for many years. This study examines current state and utility practices for collecting and reporting electricity reliability information and discusses challenges that arise in assessing reliability because of differences among these practices. The study is based primarily on reliability information for 2006 reported by 123 utilities to 37 state public utility commissions.

Chapter 4 – High winds, especially when combined with precipitation from seasonal storms, can cause damage to electricity utility systems, resulting in service interruptions to large numbers of electricity customers. While most such power outages are caused by damage from trees and tree limbs falling on local electricity distribution lines and poles, major power outages tend to be caused by damage to electricity transmission lines which carry bulk power long distances. Depending on the severity of the storm and resulting impairment, power outages can last a few hours or extend to periods of several days, and have real economic effects. Power outages can impact businesses (primarily through lost orders and damage to perishable goods and inventories), and manufacturers (mainly through downtime and lost production, or equipment damage). Data from various studies lead to cost estimates from storm-related outages to the U.S. economy at between $20 billion and $55 billion annually. Data also suggest the trend of outages from weather-related events is increasing.

Suggested solutions for reducing impacts from weather-related outages include improved tree-trimming schedules to keep rights-of-way clear, placing distribution and some transmission lines underground, implementing Smart Grid improvements to enhance power system operations and control, inclusion of more distributed generation, and changing utility maintenance practices and metrics to focus on power system reliability. However, most of these potential solutions come with high costs which must be balanced against the perceived benefits.

A number of options exist for Congress to consider which could help reduce storm-related outages. These range from improving the quality of data on storm-related outages, to a greater strategic investment in the U.S. electricity grid. Congress could empower a federal agency to develop standards for the consistent reporting of power outage data. While responsibility for the reliability of the bulk electric system is under the Federal Energy Regulatory Commission (as per the Energy Policy Act of 2005), no central responsibility exists for the reliability of distribution systems. One possible option could be to bring distribution systems under the Electric Reliability Organization for reliability purposes. Recovery after storm-related outages might be enhanced by a federal role in formalizing the review or coordination of electric utility mutual assistance agreements (MAAs). This would not necessarily mean federal approval of MAAs, but may help in the cooperative coordination of additional federal and state resources, especially in a wide, multi-state weather event. While there has been much discussion of transmission system inadequacies and inefficiencies, many distribution systems are in dire need of upgrades or repairs. The cost of upgrading the U.S. grid to meet future uses is expected to be high, with the American Society of

Civil Engineers estimating a need of $673 billion by 2020. While the federal government recently made funding available of almost $16 billion for specific Smart Grid projects and new transmission lines under the American Recovery and Reinvestment Act of 2009, there has not been a comprehensive effort to study the needs, set goals, and provide targeted funding for modernization of the U.S. grid as part of a long-term national energy strategy. Such an effort would also require decisions about the appropriate roles of government and the private sector.

Power delivery systems are most vulnerable to storms and extreme weather events. Improving the overall condition and efficiency of the power delivery system can only serve to improve the resiliency of the system, and help hasten recovery from weather-related outages. Ultimately, however, electric utilities are responsible for this infrastructure. They are in the business of selling electricity, and they cannot sell electricity if their power delivery systems are out of service.

In: Electricity Reliability in the United States ISBN: 978-1-62417-149-9
Editor: Emmanuel Martinez

Chapter 1

AN EXAMINATION OF TEMPORAL TRENDS IN ELECTRICITY RELIABILITY BASED ON REPORTS FROM U.S. ELECTRIC UTILITIES*

Joseph H. Eto, Kristina Hamachi LaCommare, Peter Larsen, Annika Todd and Emily Fisher

ABSTRACT

Since the 1960s, the U.S. electric power system has experienced a major blackout about once every 10 years. Each has been a vivid reminder of the importance society places on the continuous availability of electricity and has led to calls for changes to enhance reliability. At the root of these calls are judgments about what reliability is worth and how much should be paid to ensure it.

In principle, comprehensive information on the actual reliability of the electric power system and on how proposed changes would affect reliability ought to help inform these judgments. Yet, comprehensive, national-scale information on the reliability of the U.S. electric power system is lacking.

This chapter helps to address this information gap by assessing trends in U.S. electricity reliability based on information reported by electric utilities on power interruptions experienced by their customers. Our research augments prior investigations, which focused only on power

* This is an edited, reformatted and augmented version of Ernest Orlando Lawrence Berkeley National Laboratory publication LBNL-5268E, dated January 2012.

interruptions originating in the bulk power system, by considering interruptions originating both from the bulk power system and from within local distribution systems. Our research also accounts for differences among utility reliability reporting practices by employing statistical techniques that remove the influence of these differences on the trends that we identify.

The research analyzes up to 10 years of electricity reliability information collected from 155 U.S. electric utilities, which together account for roughly 50% of total U.S. electricity sales. The questions analyzed include:

1) Are there trends in reported electricity reliability over time?
2) How are trends in reported electricity reliability affected by the installation or upgrade of an automated outage management system?
3) How are trends in reported electricity reliability affected by the use of IEEE Standard 1366-2003?

Acronyms and Abbreviations

ASCC	Alaska Systems Coordinating Council
CAIDI	Customer Average Interruption Duration Index
CDD	cooling degree-day
EIA	Energy Information Administration
FRCC	Florida Reliability Coordinating Council
HDD	heating degree-day
HICC	Hawaiian Islands Coordinating Council
IEEE	Institute of Electrical and Electronics Engineers
MED	major event day
MRO	Midwest Reliability Organization
MWh	megawatt (10^6 watts) hour
NCDC	National Climatic Data Center
NERC	North American Electric Reliability Corporation
NOAA	National Oceanic Atmospheric Administration
NPCC	Northeast Power Coordinating Council
OMS	outage management system
RFC	Reliability First Corporation
SAIDI	System Average Interruption Duration Index
SAIFI	System Average Interruption Frequency Index
SERC	Southeast Electric Reliability Council

SPP	Southwest Power Pool
TRE	Texas Regional Entity
WECC	Western Electricity Coordinating Council

EXECUTIVE SUMMARY

Since the 1960s, the U.S. electric power system has experienced a major blackout about once every 10 years. Each has been a vivid reminder of the importance society places on the continuous availability of electricity and has led to calls for changes to enhance reliability. At the root of these calls are judgments about what reliability is worth and how much should be paid to ensure it.

The goal of this study is to inform discussions of the reliability of the U.S. electric power system by assessing trends in power interruptions experienced by U.S. electricity consumers. Our analysis is based on up to 10 years of electricity reported reliability information collected from a convenience sample of 155 U.S. electric utilities, which together account for roughly 50% of total U.S. electricity sales.

We built on prior investigations, which focused only on power interruptions originating in the bulk power system, by considering interruptions originating both from the bulk power system and from within local distribution systems. We also accounted for differences among utility practices for collecting information on and reporting power interruptions by employing statistical techniques that remove the influence of these differences on the trends we identify.

We sought to answer three questions:

1. Are there trends in reported electricity reliability over time?

We first conducted an examination relying on descriptive statistics (mean, median, customer-weighted mean) and find that reported reliability has been decreasing over time. With minor exceptions, we observed this trend for all three descriptive statistics when considering all utility reports taken together, as well as only those utility reports for which we had a complete record of 10 years of data. We point out that descriptive statistics alone mask the effects of utility-specific effects that may introduce bias into our findings.

Next, we applied rigorous statistical methods both to confirm that there were utility-specific differences among electricity reliability reports and to take explicit account of these differences in exploring correlations between reported reliability metrics and other factors. Applying these methods, we find that there are statistically significant temporal trends. We find that reported average duration and average frequency of power interruptions has been increasing over time at a rate of approximately 2% annually. In other words, reported reliability is getting worse.

While our findings are highly statistically significant, it is important to place them in appropriate context. The average annual trends we find are modest in comparison to the routinely larger year to year variations in the average duration and frequency of power interruptions experienced by utility customers. For example, in Appendix A, we present a simple analysis of trends over the most recent four years and find reported reliability has been improving over this period.

In addition, we make no claims regarding the applicability of our findings to the reliability of the U.S. electric power system as a whole. Strictly speaking, our findings apply only to the convenience sample of primarily investor-owned utilities for which we were able to collect reported reliability information. In any given year, these utilities represented roughly 50% of total U.S. electricity sales.

2. How are trends in reported electricity reliability affected by the installation or upgrade of an automated outage management system (OMS)?

A principal contribution of our work has been to examine potential sources of measurement error that could influence apparent trends in reported reliability. We find statistically significant evidence that installation or upgrade of an OMS is correlated an increase in the reported duration of power interruptions. This finding confirms anecdotal evidence long been known within the industry that reliance on prior (manual) measurement methods under-reports reliability. We also found preliminary but not statistically significant evidence for a so-called "learning effect" by which reported reliability gradually improves in years subsequent to the initial decrease in reported reliability.

Our findings might suggest that it is simply more accurate measurement of reliability, rather than lower actual reliability, which "explains" the statistically significant trend of decreasing reported reliability over time. However, our analysis takes this factor into account explicitly and still finds statistically significant secular trends toward lower reported reliability over time. Our findings, therefore, highlight the importance of taking into account the means by which reliability information is collected when examining trends in reported reliability.

3. How are trends in reported electricity reliability affected by the use of IEEE Standard 1366-2003?

We also examined a potential source of measurement bias in the form of utility reporting practices. We find that reliance on IEEE Standard 1366-2003 is correlated with higher reported reliability on average compared to reported reliability not using the IEEE standard and that this correlation is statistically significant. Nevertheless, taking this correlation into account, the secular trend of decreasing reported reliability over time remains statistically significant and at approximately the same magnitude as was found earlier (i.e., decreasing at roughly 2% annually). We caution that it is premature to attribute reliance on the IEEE standard as "causing" higher reported reliability because we could not separate the effect of reliance on the IEEE standard from other utility-specific factors (which we did not account for separately) that may also be correlated with reliance on the IEEE standard.

Next Steps

This study finds that there has been a modest, yet statistically significant secular trend of decreasing or declining reported reliability over the past 10 years. In making this finding, we summarize what our analysis to date has and has not accomplished, and outline the directions for next steps in this line of inquiry.

We wish to state clearly that, at this point, we cannot say what has caused the observed decreasing trends in reported reliability or why it is taking place. Our work has considered and characterized the influence of potential sources of measurement error or bias and found that taking these considerations into account changes neither the direction of these trends nor their statistical significance. These findings are important because they allow us to focus on potential causal factors that would help us explain the trends we observe.

To begin this process, we considered potential correlations with highly aggregated measures of weather variability and a simple measure of utility size but found neither to be statistically significant. We believe it is extremely appropriate to continue exploring differences among utilities to better understand the sources or causes of the secular trends in reliability that we observe. Some of the factors we believe should be considered include more disaggregate measures of weather variability (e.g., lightning strikes and severe storms), utility characteristics (e.g., the number of rural versus urban customers, and the extent to which transmission and distribution lines are overhead versus underground), and utility spending on transmission and distribution maintenance and upgrades, including advanced ("smart grid") technologies. It is our hope that the analysis we have conducted to date will help pave the way for these investigations and that they will be used to help ground future decisions about U.S. reliability policy, practices, and technology on a more solid factual base.

1. Introduction

Since the 1960s, the U.S. electric power system has experienced a major blackout about once every 10 years. Each has been a vivid reminder of the importance society places on the continuous availability of electricity and has led to calls for changes to enhance reliability. At the root of these calls are judgments about what reliability is worth and how much should be paid to ensure it.

In principle, information on the actual reliability of the electric power system and how proposed changes would affect reliability ought to help inform these judgments. Use of this type of information in local decision making, for example between an investor-owned utility and its state public utilities commission, is common. Yet, comprehensive, national-scale information on the reliability of the U.S. electric power system is lacking.

This chapter helps to address this information gap by assessing trends in U.S. electricity reliability based on information reported by electric utilities on power interruptions experienced by their customers. Our research augments prior investigations, which focused only on power interruptions originating in the bulk power system, by considering interruptions originating both from the bulk power and from within local distribution systems. Our research also accounts for differences among utility reliability reporting practices by employing statistical techniques that remove the influence of these differences on the trends that we identify.

The focus of prior published investigations of U.S. electric power system reliability has been primarily on the reliability of the bulk power system. For example, Amin (2008) suggests that the reliability of the bulk power system has been declining over time based on a review of the frequency and size of reported events. The response by Hines et al. (2009) rejects that hypothesis based on a rigorous statistical examination of the same data.[1]

At the same time, interruptions originating on the bulk power system represent only a small fraction of the power interruptions experienced by electricity consumers, as indicated in Hines et al. (2009) and Eto and LaCommare (2008).

The vast majority of interruptions experienced by electricity consumers are caused by events affecting primarily the electric distribution system. Thus, analyses of power interruptions originating in the bulk power system alone address only a small portion of electricity consumers' total reliability experience. Utilities routinely collect information on reliability of electric service provided to their customers. This information almost always includes all power interruptions experienced by their customers, both those originating in the bulk power system and those originating from within the electricity distribution system.

The main metrics that utilities use to report this information focus separately on the frequency and the duration of power interruptions. (See text box.)

Defining SAIDI and SAIFI

The System Average Interruption Duration Index (SAIDI) and the System Average Interruption Frequency Index (SAIFI) are metrics for the average duration and average number, respectively, of sustained power interruptions experienced by the population of customers served by a utility over the course of a year.

SAIDI and SAIFI are two of the most commonly used metrics by utilities and industry experts when reporting on the continuity of electricity service to customers.

According to IEEE Standard 1366-2003, the metrics are defined as follows:

$$\text{SAIDI} = \frac{\sum \text{Customer Interruption Durations}}{\text{Total Number of Customers Served}} \quad \text{minutes of interruption per year}$$

$$\text{SAIFI} = \frac{\sum \text{Total Number of Customers Interrupted}}{\text{Total Number of Customers Served}} \quad \text{interruptions per year}$$

Larger values of SAIDI and SAIFI indicate less reliable electricity service meaning that customers, on average, experience longer or more frequent interruptions. In this chapter, we express this relationship by describing higher or increasing reported values of SAIDI or SAIFI as an indicator of lower or declining reported reliability.

Previous work examining electric utility practices for reporting reliability information revealed significant variation (Eto and LaCommare 2008). Despite the existence of standards - albeit voluntary ones - promulgated by the industry's professional society, the Institute for Electrical and Electronics Engineers (IEEE), differences in utilities' definition and classification of power interruption events make direct comparisons among data from different utilities problematic and potentially misleading.

In this chapter, we analyze up to 10 years of reported electricity reliability information collected from a convenience sample of 155 U.S. electric utilities, which together account for roughly 50% of total U.S. electricity sales. Using these data sources, we quantify trends in electricity reliability and examine the relationship between these trends to the characteristics of the utilities, the climates in which their customers reside, utility reporting practices, and the adoption of advanced technologies for recording power interruptions. Our analysis uses statistical techniques that take into account differences in reliability reporting practices and other factors among electricity utilities, so that we can explore the effect of these differences.

The questions we examined and the motivations for examining them are as follows:

1) Are there trends in reported electricity reliability over time?

As noted above, Hines et al. (2009) concluded that there are no statistically significant trends over time based on a rigorous statistical examination of data on the reliability of the bulk power system. Taking explicit account of specific differences in utility reporting practices (and other factors) and using comprehensive information all power interruptions experienced by consumers, our analysis seeks to determine whether statistically significant temporal trends can be identified.

2) How are trends in reported electricity reliability affected by the installation or upgrade of an automated outage management system (OMS)?

McGrananghan (2006) speculated that adoption of OMS led one utility to report lower reliability because of under-reporting of customer power interruptions prior to adoption of the OMS. Our analysis explores the effect of installing or upgrading an OMS and how any such advanced reporting system is correlated with changes reported reliability over time.

3) How are trends in reported electricity reliability affected by the use of IEEE Standard 1366- 2003 (IEEE 2004)?

Eto and LaCommare (2008) compared reliability metrics reported by a convenience sample of 11 electric utilities using both historic company practices and IEEE Standard 1366-2003. Based on this small sample, those authors find no evidence of systematic bias resulting from use of the IEEE standard. The current analysis seeks to update the 2008 findings based on a larger sample of older and more recent data.

The remainder of this chapter is organized as follows:

In Section 2, we describe the information we collected to conduct the analysis, including the electricity reliability metrics, the size of the utilities reporting the metrics, the weather experienced their customers, the adoption of automated technologies for recording power interruptions, and the practices for reporting power interruptions.

In Section 3, we present findings from our preliminary investigation of time trends in reported reliability based on means, medians, and customer-weighted means.

In Section 4, we describe and present findings from application of more advanced statistical methods to the reported reliability metrics, which take into account utility-specific differences that might influence time trends in reported reliability. The utility-specific differences include the size of utility, the weather their customers experienced, installation or adoption of an automated outage management system, and utility reporting practices vis-à-vis IEEE Standard 1366-2003.

In Section 5, we summarize our main findings and discuss next steps.

Four technical appendices follow. Appendix A compares a variant of the analysis of customer-weighted means presented in Section 3, which enables a direct comparison to a closely related analysis conducted by the IEEE Distribution Reliability Working Group. Appendix B presents the results from analyses we conducted to better understand the effect of a mathematical transformation of the dependent variables examined in Section 4 prior to conducting the regression analysis. Appendix C examines the statistical outliers identified in our statistical analysis and their impact on our findings. Appendix D provides information on the results from an alternative specification of the statistical model that is the basis for findings presented in Section 4.

2. Data Collection and Review

The data we collected for this study include:

- Utility-reported reliability metrics, focusing on the System Average Interruption Frequency Index (SAIFI) and the System Average Interruption Duration Index (SAIDI),
- Installation or upgrades of automated outage management systems (OMS),
- Adoption of IEEE Standard 1366-2003 for reporting reliability metrics,
- Temperature-related weather, and
- Retail electricity sales.

This section describes the sources for these data and reviews selected aspects of the data we collected on reliability metrics.

2.1. Sources of Data

2.1.1. Utility-reported Reliability Data

Our primary source for utility-reported reliability data was state utility regulatory commissions because investor-owned utilities routinely file these data with their commissions and these data are often made publicly available (Eto and LaCommare 2008). We contacted all the commissions that made these data publicly available. As a result of this approach, the sample of utilities for which we obtained reported reliability data are largely investor-owned utilities.

In addition, we also collected some data directly from individual utilities that we had identified through previous research. No formal statistical sampling procedures were employed in determining which utilities were contacted.

Two reliability metrics, SAIDI and SAIFI, were collected for the years 2000 to 2009. We requested SAIDI and SAIFI both with and without the inclusion of major events.[2] See Section 2.2.5 for a discussion of major events and the reason why utilities sometimes report reliability metrics both including and not including these events.

We also collected information on whether and in what year a utility installed or upgraded an automated OMS. An OMS provides an automatic and consistent means for collecting information on the frequency, extent, and duration of electric service interruptions. This automation technology generally replaces manual record keeping, which is widely recognized as a less reliable means of collecting service interruption information (LET Systems 2006).

Finally, we also collected information on whether the utility relied on IEEE Standard 1366-2003 in reporting its reliability metrics. Among other things, the IEEE standard features a heuristically derived, yet systematic and statistically based method for reporting SAIDI and SAIFI without major events.

2.1.2. Temperature-related Weather Data

We collected information on weather in the form of annual heating and cooling degree-days (HDD and CDD) for 2000 to 2009 from the National Oceanic & Atmospheric Administration (NOAA) National Climatic Data Center (NCDC) (NCDC 2011).[3] HDD and CDD are measures of the need for heating or cooling in a building.

Thus, HDD is positive if ambient air temperature is cool and a building needs to be heated; CDD is positive if ambient air temperature is warm and a building needs to be cooled. HDD is defined as 65 minus the average of the daily high and low temperature where HDD is set to 0 if the average daily temperature is more than 65° F. CDD is defined as the average of the daily high and low temperature minus 65 where the CDD is set to 0 if the average daily temperature observed is less than 65°F.

We assigned state-level HDD's and CDD's to each utility based on its location.

2.1.3. Retail Electricity Sales

We collected retail electricity sales data for each utility for the years 2000 to 2009 from information that is published annually by the U.S. Energy Information Administration (EIA 2010)[4].

2.2. Review of Utility-Reported Reliability Data

2.2.1. Geographic Representation And Coverage

We collected reliability data reported by 155 different U.S. utilities. Of these, 139 are investor-owned utilities and 16 are either municipal utilities or electricity cooperatives. As noted earlier, the large number of investor-owned utilities included in our sample stems from our decision to collect their data through state public utility commissions, which routinely make these data publicly available.

Figure 1 shows the geographic distribution of these utilities by the North American Electric Reliability Corporation (NERC) region. The figure indicates, by NERC region, the number of utilities for which we collected reported reliability data and the percentage of total 2009 retail electricity sales within the region that were accounted for by these utilities.

Table 1 shows, by NERC region, the same information presented in Figure 1 as well as the percentages of total 2009 U.S. sales represented by the utilities for which we collected reported reliability data.

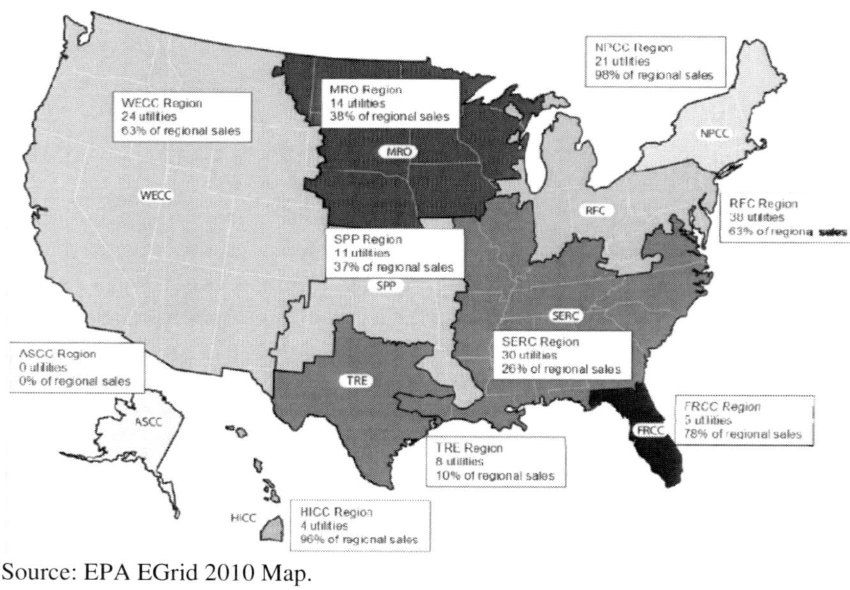

Source: EPA EGrid 2010 Map.

Figure 1. Map of U.S. by NERC Region.

Table 1. 2009 Sales of Utilities for which Data were Collected, by NERC Region

NERC Region	Total Electricity Sold in 2009 (TWh)	Total Electricity Sold by Utilities for which Data were Collected (TWh)	Percentage of Region	Percentage of U.S. Total
Western Electricity Coordinating Council (WECC)	658.7	416.4	63%	12%
Midwest Reliability Organization (MRO)	205.5	77.6	38%	2%
Southwest Power Pool (SPP)	186.1	67.9	37%	2%
Northeast Power Coordinating Council (NPCC)	222.7	218.2	98%	6%
Reliability First Corporation (RFC)	919.7	579.2	63%	16%
Southeast Electricity Reliability Council (SERC)	876.3	231.2	26%	6%
Florida Reliability Coordinating Council (FRCC)	217.9	171.0	78%	5%
Texas Regional Entity (TRE)	271.4	28.4	10%	1%
Hawaiian Islands Coordinating Council (HICC)	10.1	9.7	96%	0%
Alaska Systems Coordinating Council (ASCC)	6.3	-	0%	0%
TOTAL	3,574.7	1,799.6	50%	50%

The reliability data we collected was reported by electric utilities that together represent half of total U.S. electricity sales in 2009. The percentages of sales represented vary by region, from a low of 0% (Alaska Systems Coordinating Council [ASCC]) to a high of 98% (Northeast Power Coordinating Council [NPCC]). Reliability data from utilities representing more than 50% of total regional sales were collected for the Hawaiian Islands Coordinating Council (HICC), Florida Reliability Coordinating Council (FRCC), NPCC, Reliability First Corporation (RFC), and the Western Electricity Coordinating Council (WECC).

2.2.2. Completeness of Reported Reliability Data by Utility and Over Time

Figure 2 shows, annually from 2000 to 2009, the number of utilities whose reported reliability data we collected.

The figure shows a general increase from 2000 to 2006 in the number of utilities reporting SAIFI and SAIDI both with and without major events included.

The trend declines after 2006 for SAIFI and SAIDI without major events and by 2009 for SAIFI and SAIDI with major events.

This is likely because the most recent data were still being processed by the utilities or their regulators and were not available at the time this chapter was prepared.

Figure 3 shows the number of years of reported reliability data we collected from each of the 155 utilities. We were able to obtain a complete time series of 10 years of SAIFI and SAIDI without inclusion of major events for close to half of the utilities (70 utilities). We collected six or more years of reported reliability data for over 80% of the utilities (127 utilities).

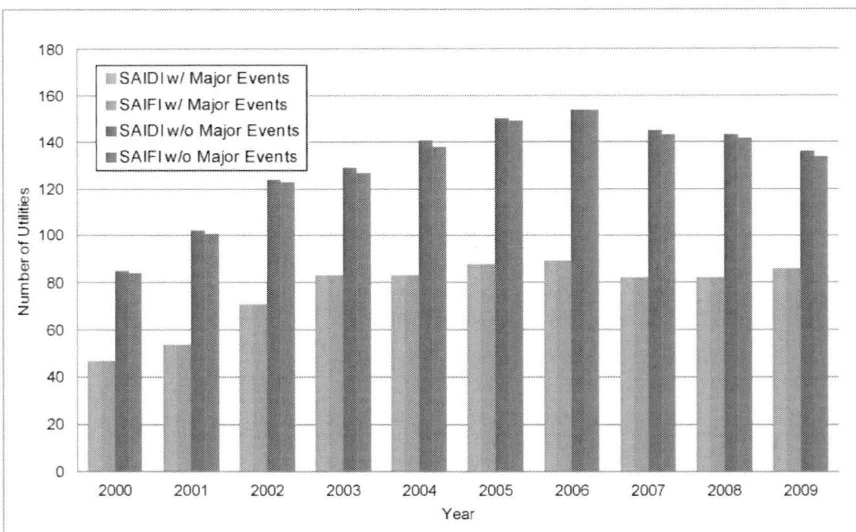

Figure 2. Number of Utilities with SAIDI and SAIFI Data.

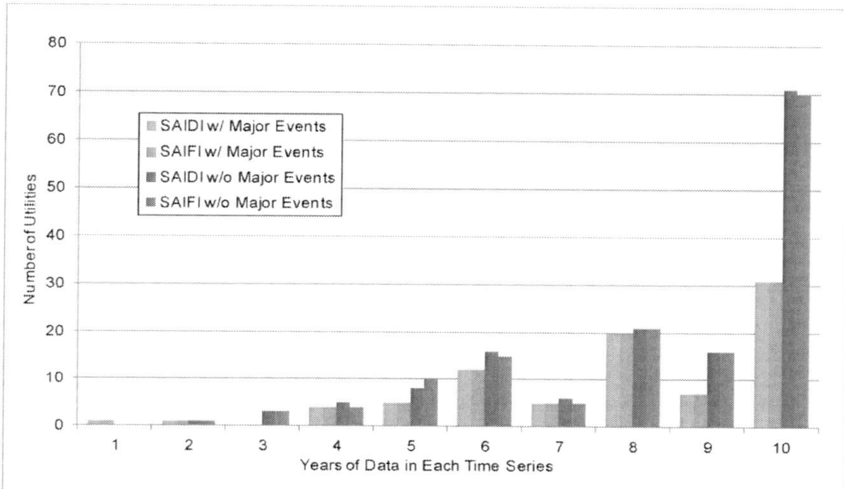

Figure 3. Completeness of Time Series.

2.2.3. Distributions of Reported SAIDI and SAIFI over Time

Figures 4 and 5 summarize, by year in box-plot form, the reported SAIDI and SAIFI values. For each year, the box-plot shows the distribution of values, both with major events (left) and without major events (right). The top and bottom of each box represent the 75th and 25th percentiles, respectively and the line through the box is the median. The mean is indicated with a blue diamond. The end points of each vertical line are the minimum and maximum values in each data set.

With the slight exception of SAIFI in year 2000, the mean values of SAIDI and SAIFI are greater when major events are included. This is to be expected. Removal of major events, which by definition are large, lowers the resulting SAIDI and SAIFI. The anomaly in the year 2000 SAIFI is due to the different mix of utilities for which we obtained SAIFI with versus without inclusion of major events. The large increase in variability in year 2008 was the result of a major hurricane.

We also examined the year to year variability in SAIDI and SAIFI for each utility. Considering the utilities for which we had a complete record of reliability metrics, we found that the mean of coefficient of variations (the ratio of the standard deviation to the mean) was more than 20% for both SAIDI and SAIFI (without major events), indicating considerable variability in the annual values of these metrics. (The means of coefficients of variation for SAIDI and SAIFI with major events were even larger.)

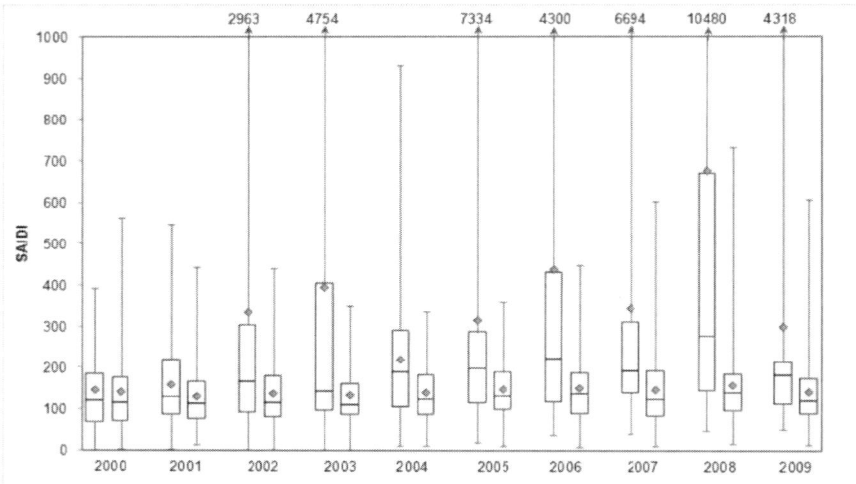

Figure 4. Box-Plot of SAIDI by Year with and without Major Events.

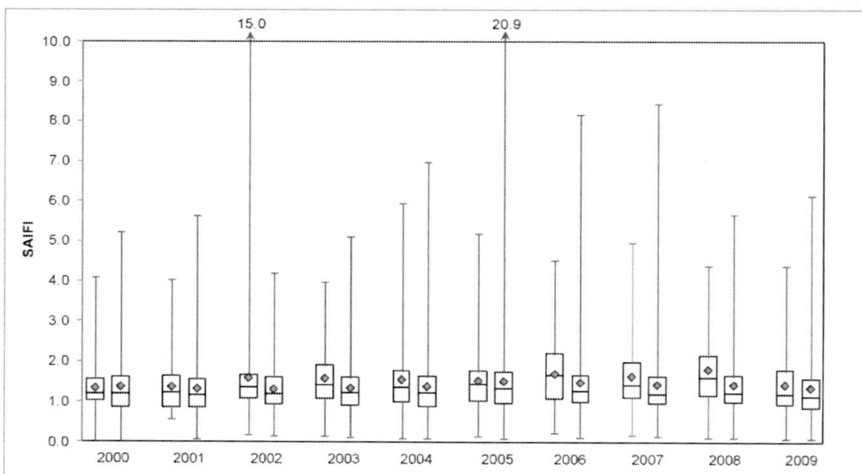

Figure 5. Box-Plot of SAIFI by Year with and without Major Events.

2.2.4. Automated Outage Management Systems (OMS)

Table 2 summarizes, by NERC region, the number of utilities that had installed or upgraded their OMS by 2010. We found that 110 utilities or 65%

of the 155 utilities for which we collected reported reliability data, had installed or upgraded their OMS by 2010.

Table 2. Summary of Utilities with an OMS

NERC Region	# Utilities We Obtained From	# Utilities that Reported They Had Installed or Upgraded their OMS by 2009
WECC	24	21
MRO	14	9
SPP	11	5
NPCC	21	16
RFC	38	24
SERC	30	16
FRCC	5	5
TRE	8	3
HICC	4	1
ASCC	0	0
TOTAL	155	100

Figure 6 presents, by NERC region, the number of utilities that installed or upgraded their OMS in each year. The line spanning years represents the cumulative number of utilities that installed or upgraded their OMS up to and including each year. The figure shows that a significant number of utilities had installed or upgraded their OMS prior to first year of our analysis (i.e., prior to 2000). The "year unknown" column in the figure represents the number of utilities that reported they had installed or upgraded their OMS, but for which we could not determine the year of installation or upgrade. In reviewing the information we collected on OMS installation or upgrade, we found that none of the utilities installed or upgraded their OMS system more than once during the 10-year time period for which we collected reported reliability data.

2.2.5. Major Events

Information on reliability is sometimes segmented using the concept of major events. Major events are extraordinary power interruptions and are defined by a variety of criteria to differentiate them from routine power interruptions.

There are a number of different definitions for major events. (See Eto and LaCommare (2008)).

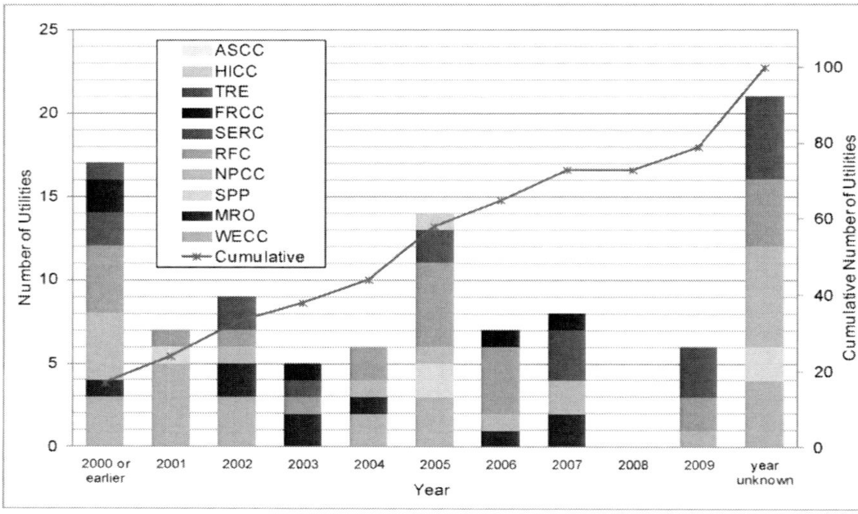

Figure 6. Number of Utilities by Year and NERC Region that Installed or Upgraded their OMS.

IEEE Standard 1366-2003 is a voluntary industry standard that articulates a consistent set of definitions and procedures for measuring and reporting distribution reliability information, including a heuristically-derived and statistically-based definition of major events.

Adoption of IEEE 1366-2003 is in its early stages. In 2006, 14 utilities (of the 120 utilities whose data we obtained) reported reliability information to their state regulatory utility commission using this standard (Eto and LaCommare 2008). For the current study, we collected reliability data for 38 utilities (of the 155 utilities whose data we collected) that reported these data using the IEEE standard.[5] (See Table 3).

The reported reliability data were prepared for all the years of data we collected either using IEEE Standard 1366-2003 or some other definition for the SAIDI and SAIFI reliability metrics.[6] In total, we collected data from 38 utilities that used the IEEE standard to report their reliability.

Of these 38 utilities, eight utilities also reported their reliability for some portion of the ten years using another set of definitions for the SAIDI and SAIFI reliability metric without major events. In preparation for the rigorous analysis of the relationship between reported reliability and reporting practices presented in section 4, we look specifically at the differences in reported

SAIDI and SAIFI without inclusion of major events, as reported by these eight utilities.

Table 3. Summary of Utilities Using IEEE 1366-2003

NERC Region	# Utilities For Which We Obtained Reported Reliability Data	# Utilities That Reported Reliability Using IEEE Std. 1366-2003
WECC	24	14
MRO	14	1
SPP	11	1
NPCC	21	4
RFC	38	12
SERC	30	6
FRCC	5	0
TRE	8	0
HICC	4	0
ASCC	0	0
TOTAL	155	38

Figures 7 presents the percentage differences between SAIDI (not including major events) reported using the IEEE standard and SAIDI (not including major events) reported using another set of definitions. Figure 8 presents the same comparison for SAIFI (not including major events). Each color represents the percentage differences for each year of data from a single utility. (Note that the utilities are not identified by name.)

Visual inspection of Figure 7 shows that SAIDI when reported using the IEEE standard is generally lower, on average, than SAIDI when reported using a method other than the IEEE standard; that is, the percentage differences are generally negative values. However, for a given utility, there is also significant variability in these values from year to year and these variations appear to be as large as, or even larger than, the average of the percentage differences over the years. Figure 8 indicates that the percentage differences for reported SAIFI using the two methods are less discernable (i.e., close to zero). In addition, the percentage variation from year to year for a given utility is also smaller, with a few notable exceptions.

It is difficult to draw definitive conclusions from such a small sample of utilities. In Section 4, we apply more sophisticated statistical methods to re-examine this topic using a much larger sample. Application of these methods will demonstrate the additional value they provide when compared to the simple comparisons presented in this section and in Section 3, which have formed the primary basis for prior analyses of these data.

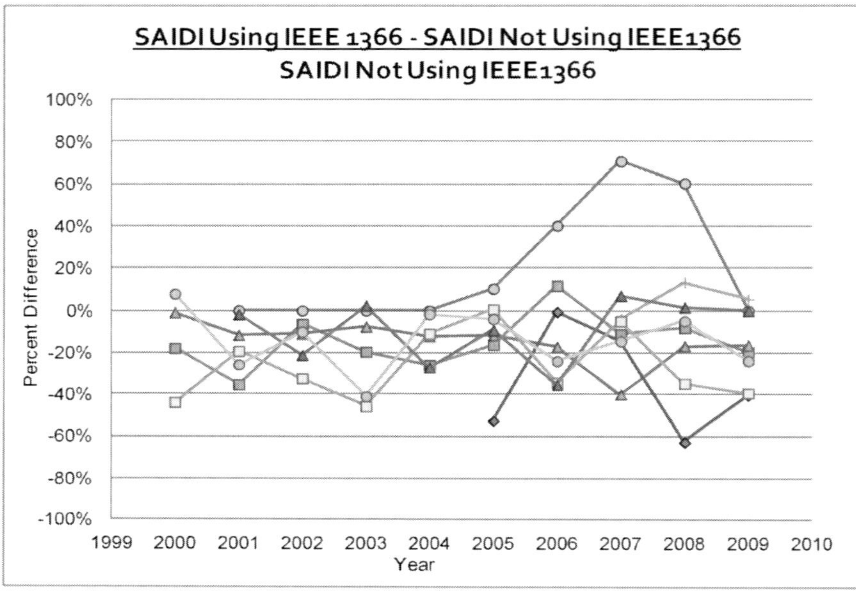

Figure 7. Percentage Difference in SAIDI Without Major Events Included between Using and Not Using IEEE Standard 1366-2003.

3. An Initial Review of Time Trends in Reported Electricity Reliability

This section presents findings from several approaches to describing time trends in reported electricity reliability. The approaches are all based on descriptive statistics, including means, medians, and customer-weighted means. The presentation compares the trends to one another and discusses considerations that affect their interpretation. Appendix A compares a variant of the analysis of customer-weighted means, which enables a direct comparison to a closely related analysis conducted by the IEEE Distribution Reliability Working Group.

Descriptive statistics cannot take into account the influences of utility-specific factors, such as the location or size of a utility, or utility-specific sources of measurement bias, such as reliance on automated outage management systems to collect reliability data or use of IEEE Standard 1366-2003 to report reliability metrics. We present a multivariate statistical analysis, which seeks to take these factors into account in Section 4.

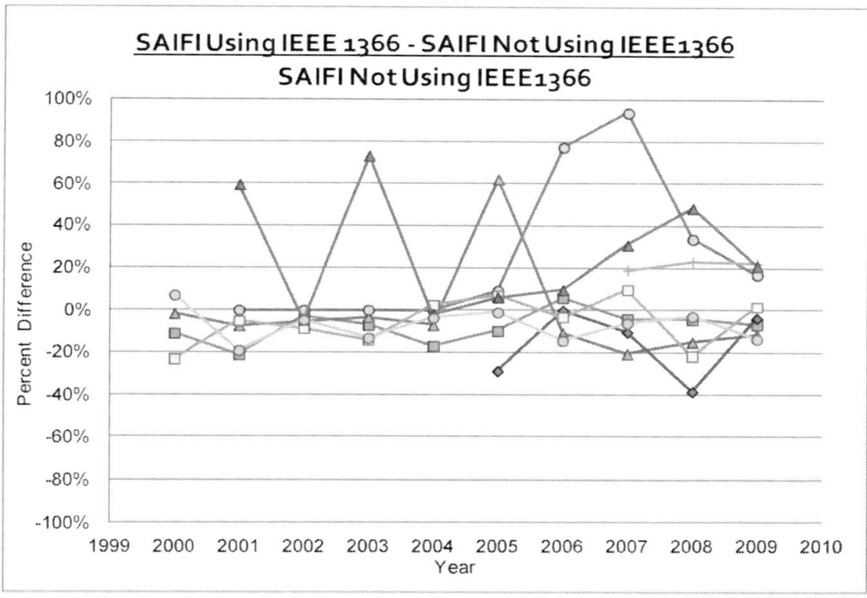

Figure 8. Percentage Difference in SAIFI Without Major Events Included between Using and Not Using IEEE Standard 1366-2003.

3.1. Time Trends in Reported Electricity Reliability Based on Descriptive Statistics

We developed time trends for each of the four reported reliability metrics (SAIDI and SAIFI both with and without inclusion of major events) using three descriptive statistics and two sets of the data we collected. The three descriptive statistics are the mean, the median, and the customer-weighted mean.

The customer-weighted mean takes into account differences in utility size and can be thought of an aggregate SAIDI and SAIFI for the entire population of included utilities, taken as a whole.

The two sets of data on reported electricity reliability are: 1) all utilities for which we had reported reliability, which we label "All;" and 2) a subset of the full set, which includes only those utilities for which we had reported reliability data for every year in the time series (years 2000-2009), which we label "Same Utilities." 7

Figures 9 through 12 plot the three descriptive statistics for SAIDI (both with and without inclusion of major events) and SAIFI (both with and without inclusion of major events), respectively, for all the utilities for which we had reported reliability data.

Figures 13 through 16 plot the three descriptive statistics for SAIDI and SAIFI (both with and without inclusion of major events), respectively, for only those utilities for which we had reported reliability data for every year in the time series (2000-2009). We had 10 years of data on SAIDI and SAIFI with major events for 28 utilities. We had 10 years of data on SAIDI and SAIFI without major events for 67 utilities.

Tables 4 and 5 present the numerical results from a best-fit linear regression for each of the trends plotted in Figures 9 through 16 for SAIDI and SAIFI, respectively.

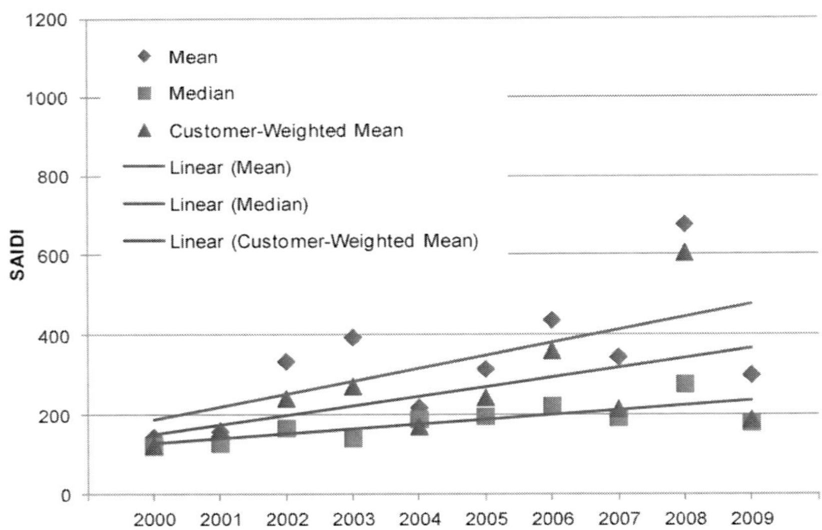

Figure 9. SAIDI with Major Events – All Reported Reliability Data.

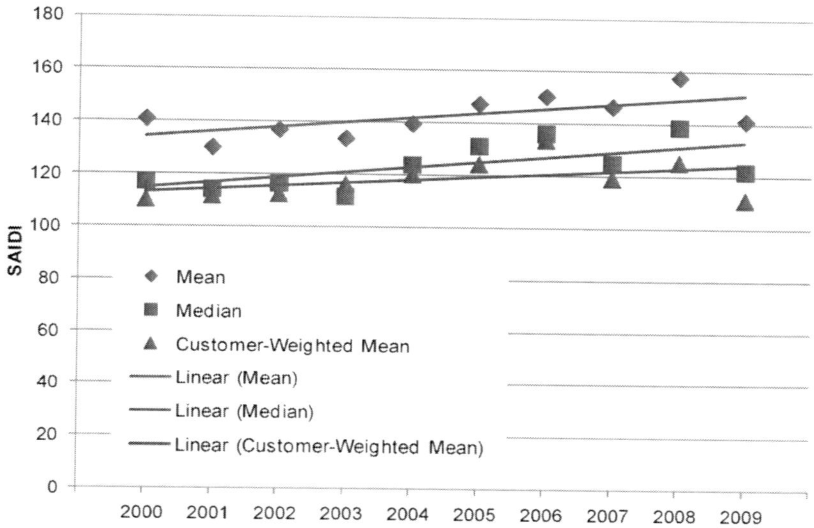

Figure 10. SAIDI without Major Events – All Reported Reliability Data.

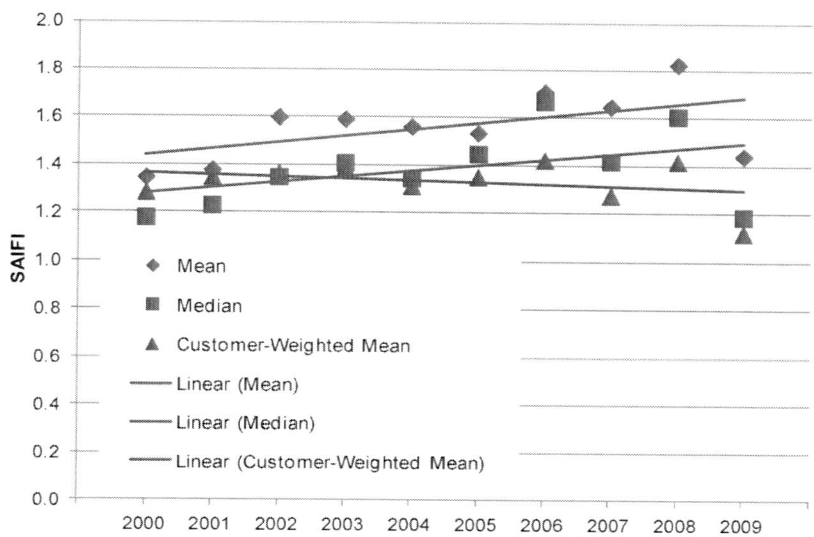

Figure 11. SAIFI with Major Events – All Reported Reliability Data.

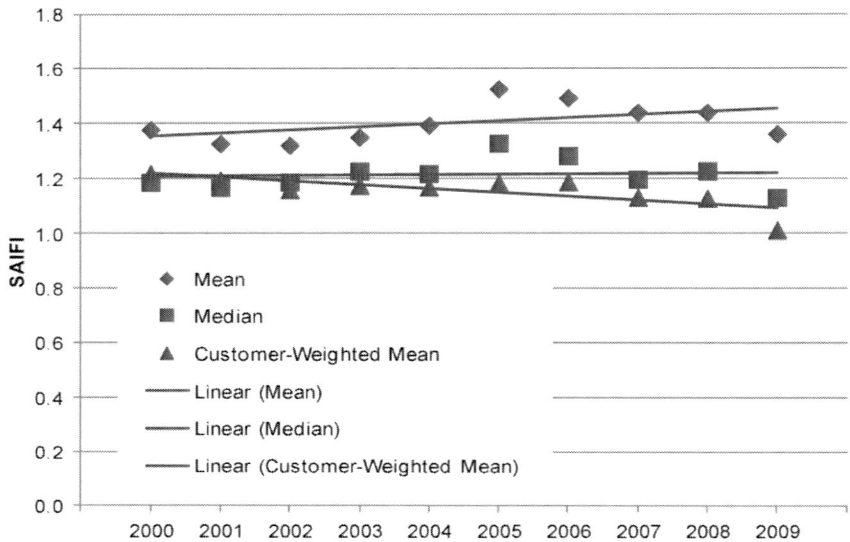

Figure 12. SAIFI without Major Events – All Reported Reliability Data.

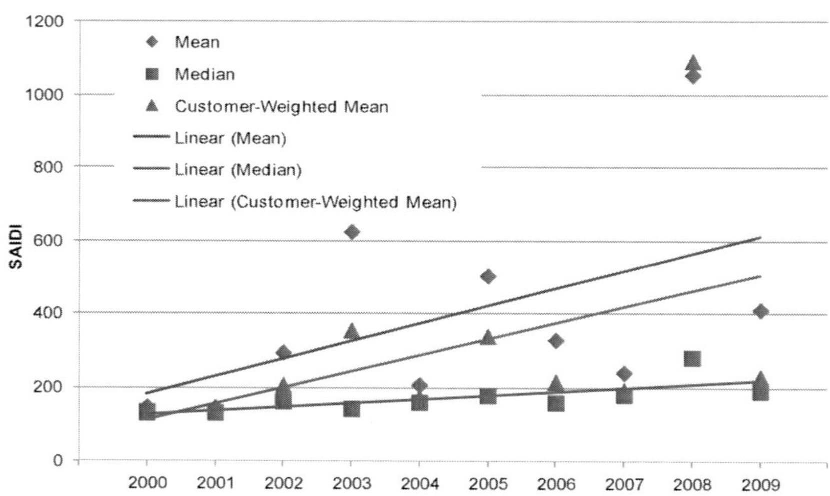

Figure 13. SAIDI with Major Events – Reported Reliability Data from Same Utilities for Every Year – N=28.

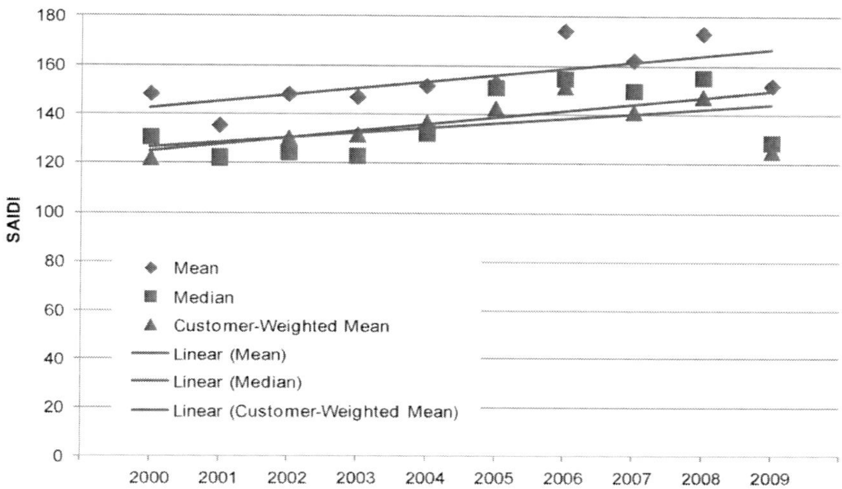

Figure 14. SAIDI without Major Events – Reported Reliability Data from Same Utilities for Every Year – N=67.

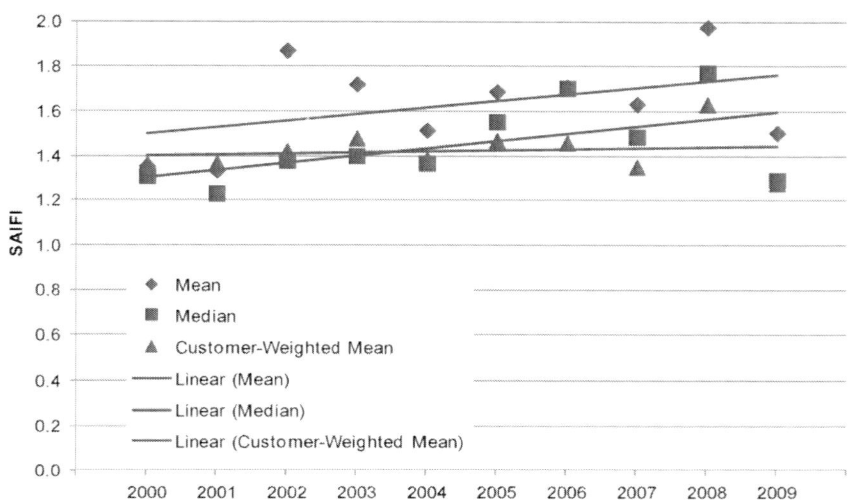

Figure 15. SAIFI with Major Events – Reported Reliability Data from Same Utilities for Every Year – N=28.

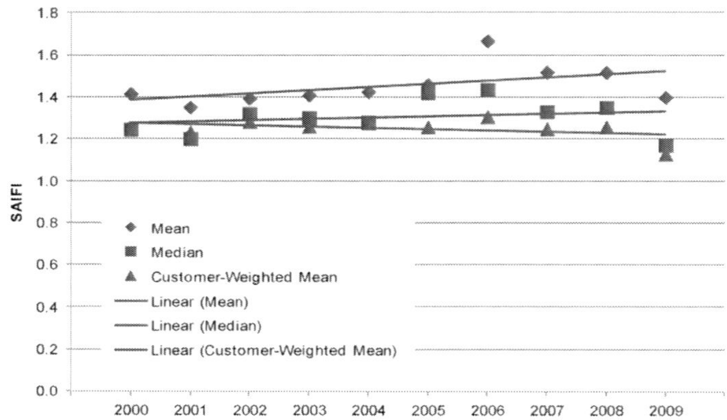

Figure 16. SAIFI without Major Events – Reported Reliability Data from Same Utilities for Every Year – N=67.

Table 4. Summary of Numerical Best Fit of Trends in SAIDI

	intercept	*slope*	*R squared*
SAIDI w/ Major Events - All Reported Reliability Data			
Average	157	32.1	0.40
Median	118	11.8	0.59
Customer Weighted Average	129	23.6	0.27
SAIDI w/o Major Events - All Reported Reliability Data			
Average	133	1.84	0.46
Median	113	2.06	0.45
Customer Weighted Average	112	1.18	0.22
SAIDI w/ Major Events - Reported Reliability Data from Same Utilities for Every Year – N=28			
Average	134	47.7	0.27
Median	116	10.4	0.53
Customer Weighted Average	69	43.6	0.21
SAIDI w/o Major Events – Reported Reliability Data from Same Utilities for Every Year – N=67			
Average	140	2.73	0.47
Median	122	2.75	0.37
Customer Weighted Average	124	1.99	0.33

* Note intercept, slope and R-squared assume x=1 (2000) to x=10 (2009)

Table 5. Summary of Numerical Best Fit of Trends in SAIFI

	intercept	*slope*	*R squared*
SAIFI w/ Major Events - All Reported Reliability Data			
Average	1.41	0.0275	0.32
Median	1.25	0.0241	0.20
Customer Weighted Average	1.37	-0.0077	0.07
SAIFI w/o Major Events - All Reported Reliability Data			
Average	1.35	0.0109	0.22
Median	1.21	0.0015	0.01
Customer Weighted Average	1.24	-0.0146	0.59
SAIFI w/ Major Events - Reported Reliability Data from Same Utilities for Every Year – N=28			
Average	1.47	0.0290	0.18
Median	1.27	0.0320	0.29
Customer Weighted Average	1.40	0.0044	0.02
SAIFI w/o Major Events – Reported Reliability Data from Same Utilities for Every Year – N=67			
Average	1.37	0.0148	0.24
Median	1.27	0.0059	0.04
Customer Weighted Average	1.28	-0.0056	0.13

* Note intercept, slope and R-squared assume x=1 (2000) to x=10 (2009).

3.2. Discussion of Time Trends Based on Descriptive Statistics

One dominant theme emerges when considering these time trends taken as a whole: the best-fit linear slopes associated with the time trends are generally positive. That is, trends in reported reliability metrics, whether assessed by considering means, medians, or customer-weighted means, indicate that the value of the metrics is increasing over time. Increasing values of SAIDI and SAIFI suggest that reported reliability is getting worse, on average, over the 10 years.[8]

This finding holds regardless of whether major events are included in the calculation of SAIDI and SAIFI. The finding also holds considering both sets of data: all reported reliability data and reported reliability from the same utilities for all 10 years.

The limited exceptions to this are the trend for customer-weighted SAIFI calculated using all reported reliability data both with and without inclusion of major event days, as well as the trend from the group of utilities for which we had a complete set of values for all 10 years (the "same" utilities) without inclusion of major event days.

Additional themes emerge from sub-groupings of these time trends.

The impacts of major events on reliability appear to have increased over time. Generally speaking, the slopes are larger for the time trends based on SAIDI and SAIFI with major events than they are for the time trends based on SAIDI and SAIFI without inclusion of major events (i.e., steeper slopes mean that reliability is getting worse faster).

On average, larger utilities, as measured by numbers of customers, would appear to be more reliable than smaller utilities. Both the intercept and slope terms are higher for the time trends based on means than they are for the time trends based on customer-weighted means.

The time trends discussed in this section are all subject to important caveats that temper the significance of these themes.

First and foremost, the statistical representativeness of the data we have collected with respect to the reliability experience of the U.S. as a whole has not been established. The findings presented in this section, while reflective of a significant portion of total U.S. electricity sales, can only be said to capture the collective reliability experience of these utilities alone and not the entire U.S. In section 2, we noted that our data are composed primarily of data from investor-owned utilities and are not drawn evenly from all regions of the U.S.

Second, the trends we examine focus on averages estimated over a period of 10 years. In Appendix A, we consider only the most recent four years of this period and find generally that, on average, reliability has improved continuously. While this reversal is not large enough to offset the overall trend for the entire 10 years, it is notable and should be acknowledged when seeking to draw conclusions regarding the significance of the overall 10-year trend.

Third, the slopes are modest in size compared the year-to-year variability that exists in the reliability metrics reported by individual utilities. In Section 2, we found that the means of the coefficients of variation (i.e., the standard deviation divided by mean) for SAIDI and SAIFI without major events from utilities for which we had all 10 years of data (67 utilities) was 20% or more. The 10-year change in values for these same reliability metrics is generally 20% or less.

Fourth, as will be examined directly in Section 4, trends based solely on descriptive statistics cannot take into account utility-specific influences that may introduce bias. Potential, yet unaccounted for, sources of bias include the means by which reliability data were collected (e.g., using an OMS versus using more manual forms of recording the frequency, extent, and duration of power interruptions), and the means by which they were reported (e.g., using IEEE Standard 1366-2003 versus individualized, state PUC-mandated reporting conventions), which are just two of several that we examine in Section 4.

4. Findings from the Statistical Analysis of Reliability Data Reported by Electric Utilities

This section describes the statistical methods we used to analyze reported electric reliability. The purpose of these methods is to take explicit account of utility-specific effects that might otherwise introduce bias into our findings. The trends presented in Section 3 were all based on descriptive statistics that cannot take these factors into account. After introducing the statistical methods, we present our findings from application of them to identify reliability trends and to correlate these trends with the factors we considered. The questions we explore in this analysis include:

1) Are there trends in reported electricity reliability over time?
2) How are trends in reported electricity reliability affected by the installation or upgrade of an OMS?
3) How are trends in reported electricity reliability affected by use of IEEE Standard 1366- 2003?

4.1. Introduction to the Statistical Methods Used in the Analysis

As described in Section 2, the reliability data we analyzed consists of up to 10 years of two reliability metrics, SAIDI and SAIFI both with and without major events, collected from up to 155 electricity distribution utilities. The data have both a cross-sectional (i.e., multiple utilities) and time-series (i.e., multiple years) element. This type of data is commonly referred to as an analysis of panel data because the methods and results involve data that have these features.

The structure and completeness of the panel data influenced our choice of tests for specifying the statistical models and the methods for estimating the model parameters and standard errors. Cameron and Trivendi (2009) refer to the specific type of panel data we analyzed as "short" because the data structure has many entities (i.e., utilities), but only a few time periods (compared to the number of entities). In addition, our panel data are unbalanced because they do not contain reliability metrics for every year from all utilities (Wooldridge 2002). In sum, our analysis is of a short, unbalanced panel data set.

The conventional statistical method used to analyze short, unbalanced panel data is multivariate regression. Multivariate regression models provide quantitative estimates of the strength of the correlation between an outcome variable (i.e., the reliability metric SAIDI or SAIFI) and a set of explanatory variables.

The specific forms of the multivariate regression models we estimated are called either "fixed effects" or "random effects" models. Fixed and random effects models are particularly useful for this type of analysis because they enable the regressions to explicitly account for differences in the outcomes (i.e., SAIDI and SAIFI) that are correlated with differences in the sources of the data for these outcomes (i.e., the utilities). As noted earlier, utilities follow different practices in reporting reliability (e.g., whether or not they use IEEE Standard 1366-2003). Fixed and random effects models can explicitly account for these correlations and thereby remove the influence of these differences from the other correlations under consideration (i.e., correlations with the other explanatory variables).

4.2. Application of the Statistical Models

Application of the statistical methods involved four sequential steps. First, we transformed the reliability metrics by expressing them as natural logarithms. Second, we conducted F-tests on the transformed reliability metrics to confirm the appropriateness of using statistical models that consider utility-specific effects. Third, we used Hausman's tests to determine whether it was more appropriate to estimate a fixed effects model versus a random effects model to capture these utility-specific effects. Fourth, we estimated two sets of models; the first consisting of a set of random effects models, and a second consisting of models that do not consider utility-specific effects. We briefly describe below each of these steps.

We decided to transform the reliability metrics by expressing them as natural logarithms for two reasons. First, it is well known that the metrics themselves tend to follow a log-normal distribution; transforming them results in a normal distribution. Second and perhaps more importantly, expression as a natural logarithm allows for a natural interpretation of the estimated coefficients from the regression equations as percentages. For example, if an estimated coefficient for an explanatory variable has a value of 0.02, the natural interpretation is that a step change in that variable correlates to a 2% increase in the reliability metric. For more information on the transformation of the data, please see Appendix B.

The F-test is a standard statistical test to determine the appropriateness of estimating fixed and random effects models. The F-test is a test of the null hypothesis that there are no fixed or random effects. If these null hypotheses can be rejected with some degree of statistical confidence, it means there may be fixed or random effects, which means the use of fixed and random effects models to estimate these effects is warranted.

We estimated two sets of log-linear models. First, we estimated a set of models that include utility-specific effects; we called this set "Model 1." The models in this set control for systematic differences across utilities, such as time, region (climate), system size, and installation or upgrade of an OMS. The set consists of separate models for SAIFI and SAIDI, both with and without inclusion of major events. We estimated both fixed and random effects versions of Model 1.

The specification of Model 1 is as follows:

$$y_{it} = \alpha + \beta_1 Sales_{it} + \beta_2 HDD_{it} + \beta_3 CDD_{it} + \beta_4 YR_t + \beta_5 OMS_{it} + \beta_6 POST_{OMSit} + \mu_i + \varepsilon_{it} \quad (1)$$

where:

y_{it} is the natural log of the reliability metric (SAIDI or SAIFI) for utility i=1,2,...,N in year
t=1,2,...,T;
$Sales_{it}$ is annual electricity sales in Millions of MWh;
HDD_{it} is heating degree-days;
CDD_{it} is cooling degree-days;
YR_t is a time trend in years;

OMS_{it} is an indicator variable that takes the value 1 if utility *i* has an OMS installed in year *t*, and 0 otherwise

$POST_{OMSit}$ takes the value 1 for the first year after utility *i* installs an OMS, 2 for the second year after an OMS is installed, etc, and 0 for earlier years prior to the installation of an OMS; and p_i is the utility-specific error.

We applied the Hausman (1978) specification test to Model 1 to determine whether the fixed or random effects version of this model was more appropriate. The Hausman test examines whether, under the null hypothesis, the individual utility effects are uncorrelated with the other regressors in the model. If the null hypothesis is not rejected, both the random effects and the fixed effects models are consistent, but only the random effects model is efficient. This means that fixed and random effects models will have the same expected values, but the random effects model will have much smaller standard errors. Using a fixed effects model when the random effects model is consistent may lead to an erroneous interpretation of the statistical significance of coefficients. See Greene (2000) for a more detailed discussion of the difference between fixed and random effects. The Hausman test did not reject the null hypothesis of random effects in six of the eight regressions we ran (see Tables 7 and 8). We therefore concluded–in general–that the random effects model was consistent and more efficient than the fixed effects version.

Second, we estimated a set of models that did not include utility-specific effects, which we called "Model 2." The models sought to examine how reporting differences, specifically utilization of IEEE Standard 1366-2003, along with other unobserved correlates with utilization of the IEEE standard, correlate with reported reliability. The specification of Model 2, which does not include utility-specific effects, p_i, is as follows:[9]

$$y_{it} = \alpha + \beta 1 IEEE_i + \beta 2 Sales_{it} + \beta 3 HDD_{it} + \beta 4 CDD_{it} + \beta 5 YR_t + \beta 6 OMS_{it} + \beta 7 POST_{OMSit} + \epsilon_{it} \quad (2)$$

where, in addition to the variables defined above:

$IEEE_i$ takes the value 1 if utility *i* reports interruptions using IEEE Standard 1366-2003, and 0 otherwise.

Note that for both Model 1 and Model 2, the time trend, YR_t, enters as a linear time variable rather than as a year-specific effect. This additional assumption was deemed reasonable because the cost of including it as a year-specific effect is high, in terms of degrees of freedom, for such a short dataset.[10] See Appendix D for a model that includes the time trend as a year-specific effect; the time trend is similar, but each individual year is not statistically significant.

We estimated standard errors for the one-way unbalanced data model using a specialization (Baltagi and Chang 1994) of the approach proposed by Wansbeek and Kapteyn (1989) for unbalanced two-way models. The Wansbeek and Kapteyn method for estimating variance components is the default approach used by SAS in the one-way random effects estimation of unbalanced panel data (SAS 2011b).

4.3. Findings

4.3.1. Are There Utility-Specific Differences in Reported Electricity Reliability?

Table 6 presents the results from the application of the F-test to the reliability metrics. The table indicates that both one-way (utility only) and two-way (utility and year) effects are statistically significant (at the 0.01% confidence level) for all four reliability metrics – SAIDI and SAIFI both with and without major events. That is, there are very strong correlations between the utility and the values of the reliability metrics as well as between the utility and the year when correlated to the values of the reliability metrics.

The strong correlation between the utility alone and the reliability metrics indicates that it is important to take this correlation into account when examining correlations between the reliability metrics and other correlated (or explanatory) variables. In other words, there are strong utility-specific effects that are systematically correlated with the reliability metrics.

At this point, we cannot determine the exact or complete set of sources or causes of these effects, but they are consistent with the existence of utility-specific differences in reporting practices. Hence, taking this correlation into account appropriately means that subsequent correlations with other variables will not be "contaminated" by these differences in reporting practices (by any other utility-specific effects).

The correlation between utility plus year to the reliability metrics means that year-to-year correlations are also important to take into account when examining correlations with other variables.

This finding supports examining the reliability metrics, by utility, as a time series, rather than as a handful of observations randomly drawn from different years.

Table 6. F-test of the Hypothesis that there are No Utility-Specific Effects

	One-way Fixed Effect (Utility)			Two-way Fixed Effects (Utility and Year)		
Reliability Metric	F Value	Degrees of Freedom (among/within)	Prob. > F	F Value	Degrees of Freedom (among/within)	Prob. > F
ln SAIDI (w/o MEs)	15.29	143/1037	< 0.0001	14.80	152/1029	< 0.0001
ln SAIFI (w/o MEs)	15.67	143/1034	< 0.0001	15.08	152/1026	< 0.0001
ln SAIDI (w/MEs)	5.32	85/595	< 0.0001	5.74	94/587	< 0.0001
ln SAIFI (w/MEs)	9.61	85/595	< 0.0001	9.73	94/587	< 0.0001

Note: ME = major event, ln = natural logarithm,

Note: The SAS software test for effects of cross-level interactions (utility or utility and year) reports two types of degrees of freedom: 1) "among" and 2) "within". The "among" value is equal to k-1 degrees of freedom where k is the number of cross-sections per effect. The "within" value is equal to N-k degrees of freedom, where N is the total number of observations and k is the number of cross-sections per effect.

The bottom two rows of Table 7 include the results from applying the Hausman test to Model 1. The test does not reject the null hypothesis. We therefore conclude that a random effects model is consistent and more efficient than the fixed effects version in this case.[11] Accordingly, we present results for the random effects model only.[12]

4.3.2. Are there Trends in Reported Electricity Reliability Over Time?

We estimated each model separately for each of the four different reliability metrics: SAIFI and SAIDI, both with and without major events. We estimated Model 1 with two specifications for the treatment of installation or upgrade of an OMS. The first version considers only the differences in reported reliability before and after installation or upgrade. The second version considers a "learning" effect in which the model estimates the correlation with installation or upgrade in subsequent years.

Tables 7 and 8 present the results for the Model 1. Table 7 presents the results for SAIFI and SAIDI with and without major events for the initial version of Model 1 (i.e., without OMS learning). Table 8 presents the results for SAIFI and SAIDI with and without major events for the second versions of Model 1 (i.e., with OMS learning).

Table 7. One-way Random Effects Regression (Model 1): The Effect of Sales, HDD, CDD, Time, and OMS on Frequency and Duration of Interruptions (with Major Events Included)

	With Major Events Included				Without Major Events Included			
	I ln SAIFI		II ln SAIDI		III ln SAIFI		IV ln SAIDI	
Intercept	-42.6706	***	-89.3893	***	-35.9339	***	-39.1642	***
	(13.3111)		*(25.7755)*		*(8.4430)*		*(9.3735)*	
Sales	-0.00153		-0.00133		-0.00112		-0.00246	
	(0.0023)		*(0.0035)*		*(0.0020)*		*(0.0023)*	
HDD	-0.00002		6.65E-06		-0.00002		-0.00003	
	0.0000		*(0.0001)*		*0.0000*		*0.0000*	
CDD	-7.01E-06		-0.00006		0.000133	**	0.000072	
	(0.0001)		*(0.0001)*		*(0.0001)*		*(0.0001)*	
YR	0.021496	***	0.04716	***	0.01797	***	0.02194	***
	(0.0067)		*(0.0129)*		*(0.0042)*		*(0.0047)*	
OMS	0.004575		0.287343	***	-0.04346		0.136875	***
	(0.0561)		*(0.1020)*		*(0.0404)*		*(0.0452)*	
POST OMS								
Utility Effects	Yes		Yes		Yes		Yes	
R-square	0.0205		0.0581		0.0265		0.0632	
Hausman								
Test (m Value)	3.21		2.49		2.22		1.75	
Hausman	Fail to Reject		Fail to Reject		Fail to Reject		Fail to Reject	
Interpretation	Null		Null		Null		Null	

Note: A generalization of the R-square measure is reported and is based on Buse (1973). This generalized goodness-of-fit measure is the proportion of the transformed sum of squares of the dependent variable that is attributable to the influence of the independent variables exclusive of utility-specific random effects. Hausman (1978) tests the null hypothesis that random effects are preferred over fixed effects. Standard errors are in parentheses. * $p<0.10$, ** $p<0.05$, *** $p<0.01$.

Both tables show evidence of a secular trend of increasing frequency and duration of interruptions on average over the years 2000-2009. [13] In Table 7, SAIFI and SAIDI without major events, columns III and IV, the coefficients for *YR* are 0.018 and 0.022 for SAIFI and SAIDI, respectively. Both of these estimated coefficients are statistically significant at the 1% confidence level.

The natural interpretation of these coefficients is that SAIFI and SAIDI are increasing annually, by about 2% for both SAIFI and SAIDI.

Table 8. One-way Random Effects Regression (Model 1): The Effect of Sales, HDD, CDD, Time, OMS, and OMS "Learning" on Frequency and Duration of Interruptions (Without Major Events Included)

	With Major Events Included				Without Major Events Included			
	I ln SAIFI		II ln SAIDI		III ln SAIFI		IV ln SAIDI	
Intercept	-65.2081	***	-125.533	***	-42.9671	***	-46.763	***
	(17.1806)		*(31.8647)*		*(9.7397)*		*(10.8662)*	
Sales	-0.00114		-0.00081		-0.00087		-0.00221	
	(0.0023)		*(0.0035)*		*(0.0020)*		*(0.0023)*	
HDD	-0.00002		2.69E-07		-0.00002		-0.00003	
	0.0000		*(0.0001)*		*0.0000*		*0.0000*	
CDD	-0.00002		-0.00009		0.000131	**	0.00007	
	(0.0001)		*(0.0001)*		*(0.0001)*		*(0.0001)*	
YR	0.032765	***	0.065237	***	0.021482	***	0.025734	***
	(0.0086)		*(0.0159)*		*(0.0049)*		*(0.0054)*	
OMS	0.044586		0.380678	***	-0.01665		0.163991	***
	(0.0593)		*(0.1129)*		*(0.0444)*		*(0.0493)*	
POST OMS	-0.02411	**	-0.04141	*	-0.01257		-0.01337	
	(0.0117)		*(0.0216)*		*(0.0087)*		*(0.0097)*	
Utility Effects	Yes		Yes		Yes		Yes	
R-square	0.0292		0.0477		0.031		0.0491	
Hausman Test (m Value)	4.76		11.54	**	5.68		10.78	*
Hausman	Fail to Reject				Fail to Reject			
Interpretation	Null		Reject Null		Null		Reject Null	

Notes: A generalization of the R-square measure is reported and is based on Buse (1973). This generalized goodness-of-fit measure is the proportion of the transformed sum of squares of the dependent variable that is attributable to the influence of the independent variables exclusive of utility-specific random effects. Hausman (1978) tests the null hypothesis that random effects are preferred over fixed effects. Standard errors are in parentheses. * $p<0.10$, ** $p<0.05$, *** $p<0.01$.

It is useful to note that 2% annual decreases in reported reliability are roughly consistent with the simple linear trends presented in Tables 4 and 5 (and in Figures 9 through 16) in section 3. In this regard, the observation first made in section 3 – that these trends are modest in comparison to the year-to-year variability in these reported reliability metrics – also apply equally to these findings.

These trends are also confirmed when major events are not included in SAIFI and SAIDI.

In Table 7, columns I and II, the coefficients for *YR* are 0.022 and 0.047 for SAIFI and SAIDI, respectively. Again both are statistically significant at the 1% level. The natural interpretation of these coefficients is that SAIFI is increasing annually at about 2% and SAIDI is increasing annually at about 5%.

It is important to observe that, in making these estimates, Model 1 also takes into account the potential for correlations with electricity sales and climate (as well as the effect of utility-specific differences, as discussed in Section 4.3.1).

In fact, the *model* finds that these external factors are generally not at all correlated with the increasing secular trends observed for both SAIFI and SAIDI.

One exception is that correlation between SAIFI without major events and CDDs is statistically significant at the 5% level. The natural interpretation of this correlation is that SAIFI is very slightly higher when there are more CDDs.

It is premature to speculate or draw conclusions about the causes underlying these trends without more explicit treatment of potential sources of bias in reported reliability.[14] In the next subsection, we focus on measurement error as one potential source of bias.

4.3.3. How are Trends in Reported Electricity Reliability Affected by the Installation or Upgrade of The Automated OMS?

The estimates for the *OMS* coefficients in Table 7, describe the strength of the correlation between installation or upgrade of an OMS and reported SAIFI and SAIDI. The results differ for SAIFI and SAIDI, both in terms of the direction and strength of the correlation.

The correlation of SAIFI (both with and without major events) to the installation or upgrade of an OMS is mixed and not statistically significant, even at a 10% confidence level. However, the correlation with SAIDI is always positive and is statistically significant at the 1% confidence level. The natural interpretation of this correlation is that utilities that install or upgrade their OMS report higher SAIDI by nearly 29% when major events are included and by nearly 14% when major events are not included compared to utilities that did not install or upgrade their OMS.

Table 8 further explores the relationship between installation or upgrade of an automated OMS and reported reliability by introducing a time-element, *POST OMS*, which tracks how the correlation with SAIFI and SAIDI changes in the years following installation or upgrade of the system.

The results are suggestive, but not conclusive. The results are suggestive because there is evidence that installation or upgrade of an OMS is correlated with an initial increase in SAIFI or SAIDI, but that SAIFI and SAIDI decrease in the years following installation or upgrade. The results are not conclusive because the estimated coefficients are not consistently statistically significant.

The only SAIFI coefficient that is statistically significant is *POST OMS* with major events included, at the 5% confidence level. The natural interpretation is that there is an annual reduction in SAIFI of 2.4% following installation or upgrade of an OMS compared to the SAIFI reported by utilities that did not install or upgrade their OMS.

The SAIDI coefficients that are statistically significant include *OMS* (1% level) and *POST OMS* (10% level) with major events included, and *OMS* (1% level) when major events are not included. The natural interpretations are that when major events are included, there is a one-time increase in SAIDI of 38% followed by an annual decrease of 4%, compared to SAIDI reported utilities that did not install or upgrade their OMS. When major events are not included, there is a one-time increase in SAIDI of 16%.

The effects of installation or upgrade of an OMS in the years following installation or upgrade are not consistently statistically significant. However, the coefficients on the magnitude of the year-to-year changes in the reliability metrics remain highly statistically significant at the 1% level. In fact, SAIFI and SAIDI with major events increase annually at faster rates (of about 3% and 6.5%, respectively) than the estimates that do not consider this effect. SAIFI and SAIDI without major events increase annually at rates of about 2% and 2.5%, respectively, which is roughly consistent with the estimates that do not consider this effect.[15]

4.3.4. How are Trends In Reported Electricity Reliability Affected by the Use of IEEE Standard 1366-2003?

Table 9 presents the results for Model 2, which removes γ_i, the company-specific effect, and replaces it with $IEEE_i$, which indicates whether the company relied on IEEE Standard 1366-2003 in reporting its reliability metrics.

Table 9 was developed for only SAIFI and SAIDI *without major events* because reliance on IEEE Standard 1366-2003 involves implementing a specific method for not including these events.

Table 9 reports that the coefficient for reliance on IEEE Standard 1366-2003 is not statistically significant for SAIFI and is statistically significant at the 5% level for SAIDI.

Table 9. No Utility Fixed Effects Regression (Model 2): Effect of IEEE, sales, HDD, CDD, Time, and OMS on Frequency and Duration of Interruptions (without major events)

	ln SAIFI		ln SAIDI	
Intercept	-34.5795	**	-43.9178	***
	(13.5039)		*(14.9314)*	
IEEE	0.023445		-0.10797	**
	(0.0425)		*(0.0470)*	
Sales	-0.00226	**	-0.00446	
	(0.0010)		*(0.0011)*	
HDD	-0.00002		-0.00008	
	(0.0000)		*(0.0000)*	
CDD	0.000109	***	-0.00011	***
	(0.0000)		*(0.0000)*	
YR	0.01734	**	0.024603	***
	(0.0068)		*(0.0075)*	
OMS	-0.00033		0.041122	
	(0.0602)		*(0.0663)*	
POST OMS	-0.01839		-0.01123	
	(0.0114)		*(0.0126)*	
Utility Fixed Effects	No		No	

Notes: Standard errors in parentheses. * $p<0.10$, ** $p<0.05$, *** $p<0.01$.

The natural interpretation of the latter result is that reliance on IEEE Standard 1366-2003 is correlated with a lower reported SAIDI of about 11%.

However, we caution the reader that, strictly speaking, this interpretation is premature. The most that can be said is that reliance on the IEEE standard, *along with all other utility-specific effects that are highly correlated with reliance on the IEEE standard*, is correlated with reported reliability in this manner.

We leave it to future work to develop specifications that would separate the effect of reliance on the IEEE standards from these other correlates to isolate the impact of this effect uniquely.

The removal of utility-specific effects also affects the values and statistical significance of other correlates in the model, compared to the values estimated for them in Model 1. For SAIFI, the coefficients on *Sales* and *CDD* are also statistically significant in Model 2. For SAIDI, the coefficient on *CDD* is statistically significant in Model 2.

The statistical significance of these correlates is likely reflective of a utility-specific effect because these correlations were not at all or less statistically significant when utility-specific effects were taken into account (in Model 1).

In contrast, the time trend, *YR*, for both SAIFI and SAIDI is, like Model 1, statistically significant.

The natural interpretation is that SAIFI and SAIDI are increasing at slightly less than 2% and nearly 2.5% annually, which is roughly consistent with the interpretation of the coefficients for Model 1 presented in Table 8.

5. Summary and Interpretation of Findings, and Next Steps

The goal of this study is to inform discussions of the reliability of the U.S. electric power system by assessing trends in power interruptions experienced by U.S. electricity consumers. Our analysis is based on up to 10 years of electricity reported reliability information collected from a convenience sample of 155 U.S. electric utilities, which together account for roughly 50% of total U.S. electricity sales.

We built on prior investigations, which focused only on power interruptions originating in the bulk power system, by considering interruptions originating both from the bulk power system and from within local distribution systems. We also accounted for differences among utility practices for collecting information on and reporting power interruptions by employing statistical techniques that remove the influence of these differences on the trends we identify.

We sought to answer three questions:

1) Are there trends in reported electricity reliability over time?

We first conducted an examination relying on descriptive statistics (mean, median, customer-weighted mean) and find that reported reliability has been decreasing over time. With minor exceptions, we observed this trend for all three descriptive statistics when considering all utility reports taken together, as well as only those utility reports for which we had a complete record of 10 years of data. We point out that descriptive statistics alone mask the effects of utility-specific effects that may introduce bias into our findings.

Next, we applied rigorous statistical methods both to confirm that there were utility-specific differences among electricity reliability reports and to take explicit account of these differences in exploring correlations between reported reliability metrics and other factors. Applying these methods, we find that there are statistically significant temporal trends. We find that reported average duration and average frequency of power interruptions has been increasing over time at a rate of approximately 2% annually. In other words, reported reliability is getting worse.

While our findings are highly statistically significant, it is important to place them in appropriate context. The average annual trends we find are modest in comparison to the routinely larger year to year variations in the average duration and frequency of power interruptions experienced by utility customers. For example, in Appendix A, we present a simple analysis of trends over the most recent four years and find reported reliability has been improving over this period.

In addition, we make no claims regarding the applicability of our findings to the reliability of the U.S. electric power system as a whole. Strictly speaking, our findings apply only to the convenience sample of primarily investor-owned utilities for which we were able to collect reported reliability information. In any given year, these utilities represented roughly 50% of total U.S. electricity sales.

2) How are trends in reported electricity reliability affected by the installation or upgrade of an automated outage management system (OMS)?

A principal contribution of our work has been to examine potential sources of measurement error that could influence apparent trends in reported reliability.

We find statistically significant evidence that installation or upgrade of an OMS is correlated an increase in the reported duration of power interruptions.

This finding confirms anecdotal evidence long been known within the industry that reliance on prior (manual) measurement methods under-reports reliability.

We also found preliminary but not statistically significant evidence for a so-called "learning effect" by which reported reliability gradually improves in years subsequent to the initial decrease in reported reliability.

Our findings might suggest that it is simply more accurate measurement of reliability, rather than lower actual reliability, which "explains" the statistically significant trend of decreasing reported reliability over time. However, our analysis takes this factor into account explicitly and still finds statistically significant secular trends toward lower reported reliability over time. Our findings, therefore, highlight the importance of taking into account the means by which reliability information is collected when examining trends in reported reliability.

3) How are trends in reported electricity reliability affected by the use of IEEE Standard 1366-2003?

We also examined a potential source of measurement bias in the form of utility reporting practices. We find that reliance on IEEE Standard 1366-2003 is correlated with higher reported reliability on average compared to reported reliability not using the IEEE standard and that this correlation is statistically significant.

Nevertheless, taking this correlation into account, the secular trend of decreasing reported reliability over time remains statistically significant and at approximately the same magnitude as was found earlier (i.e., decreasing at roughly 2% annually).

We caution that it is premature to attribute reliance on the IEEE standard as "causing" higher reported reliability because we could not separate the effect of reliance on the IEEE standard from other utility-specific factors (which we did not account for separately) that may also be correlated with reliance on the IEEE standard.

Next Steps

This study finds that there has been a modest, yet statistically significant secular trend of decreasing or declining reported reliability over the past 10 years. In making this finding, we summarize what our analysis to date has and has not accomplished, and outline the directions for next steps in this line of inquiry.

We wish to state clearly that, at this point, we cannot say what has caused the observed decreasing trends in reported reliability or why it is taking place. Our work has considered and characterized the influence of potential sources of measurement error or bias and found that taking these considerations into account changes neither the direction of these trends nor their statistical significance. These findings are important because they allow us to focus on potential causal factors that would help us explain the trends we observe.

To begin this process, we considered potential correlations with highly aggregated measures of weather variability and a simple measure of utility size but found neither to be statistically significant. . However, these examinations are preliminary and far from complete. For example, with respect to the influence of weather variability, we can only conclude that annual HDDs and CDDs as a measure of yearly weather are not well-correlated with the reported reliability metrics. However, these are only two measures of yearly weather variability; there are others that could be studied. Similarly, utility size is only one measure of the many potential differences among utilities that might be correlated with reported reliability.

We believe it is extremely appropriate to continue exploring differences among utilities to better understand the sources or causes of the secular trends in reliability that we observe. Some of the factors we believe should be considered include more disaggregate measures of weather variability (e.g., lightning strikes and severe storms), utility characteristics (e.g., the number of rural versus urban customers, and the extent to which transmission and distribution lines are overhead versus underground), and utility spending on transmission and distribution maintenance and upgrades, including advanced ("smart grid") technologies.

It is our hope that the analysis we have conducted to date will help pave the way for these investigations and that they will be used to help ground future decisions about U.S. reliability policy, practices, and technology on a more solid factual base.

Appendix A. Customer Weighted Average Comparison to IEEE DRWG Benchmarking Analysis

The IEEE Distribution Reliability Working Group (DRWG) conducts an annual benchmarking analysis based on reliability metrics that are submitted on a voluntary basis. At the IEEE Summer 2011 General Meeting, the DRWG presented a customer-weighted time trend that we can compare to a variant of the customer-weighted time trend presented in Section 3. We made several adjustments to the customer-weighted time trend presented in Section 3 in order to facilitate a more direct comparison with the time trends developed by the DRWG. First, we compare only the years 2006 through 2009, which are the same years for which the DRWG developed its time trend. Second, we compare only those utilities that relied on IEEE Standard 1366-2003 to report their reliability metrics, which are the only utilities that DRWG considers in developing its time trend. Third, we compare only SAIDI and SAIFI without inclusion of major events, again, to be consistent with DRWG. Fourth, we develop our trends based only on those utilities for which we had all four years of reported reliability, again, to be consistent with DRWG. We label the adjusted customer-weighted time trends "Revised LBNL"

Figures A-1 and A-2 present both the original customer-weighted means and trend line best fit equations for all reported reliability data from Section 3 along with the DRWG's and the revised LBNL customer-weighted means for SAIDI without inclusion of major events and SAIFI without inclusion of major events, respectively.

We find that, the DRWG and the revised LBNL trend lines for both SAIDI and SAIFI are consistent with one another. Both are downward sloping, indicating that over the period 2006- 2009, reported reliability, on a customer-weighted basis, has been improving (i.e., reported reliability metrics indicate that reliability is getting better).

This finding contrasts with the time trends presented in Section 3, which found that over the ten-year period from 2000 to 2009, reported reliability metrics were generally increasing (i.e., reliability was getting worse) over time.

As a reminder, the same caveats applied to the time trends presented in Section 3 also apply to the findings presented in this Appendix. First, neither the statistical representativeness of the samples of reported reliability data examined by LBNL nor those included in the DRWG 2011 Benchmarking

Survey compared to the U.S as a whole have been established. Second, the influence of utility-specific effects, such as biases that may have been introduced by the decision to report reliability data following IEEE Standard 1366-2003 and reliance on automated outage management system to collect reliability data, among other unexamined sources of potential bias, are not taken into account.

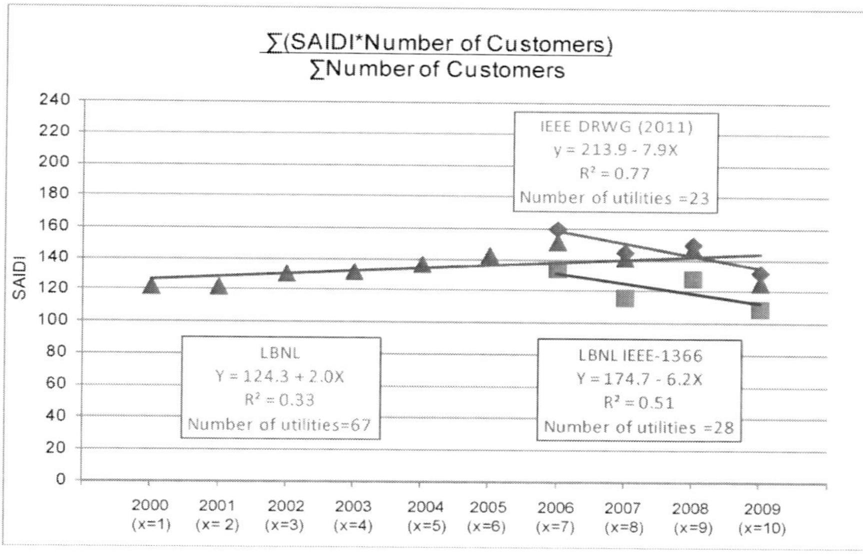

Figure A- 1. Customer-weighted SAIDI w/o Major Events – Comparison of LBNL Findings to those of the IEEE DRWG 2011 Benchmarking Survey.

APPENDIX B. WHY A LOG-NORMAL DISTRIBUTION?

Section 4.2 describes, among other things, our decision to transform the reliability metrics (annual SAIDI and SAIFI) examined in the regression analysis by expressing them as natural logarithms. This appendix describes and documents the rationale for this decision. The rationales all involve describing why we found it desirable to utilize the reliability metrics re-expressed as natural logarithms rather than utilize them un-transformed.

To summarize, we decided to transform the dependent variables for three reasons.

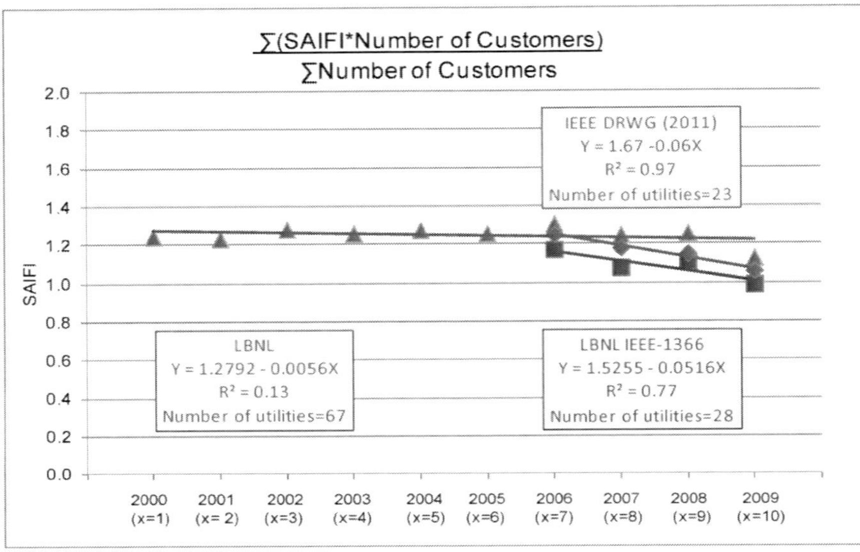

Figure A- 2. Customer-weighted SAIFI w/o Major Events – Comparison of LBNL Findings to those of the IEEE DRWG 2011 Benchmarking Survey.

First, we found encouraging visual evidence that, when expressed as natural logarithms, the distribution of annual values of SAIDI and SAIFI we collected appeared to follow a normal distribution better than the un-transformed annual values. Second, we performed statistical tests that gave a positive indication that expression of SAIDI without inclusion of major events as a natural logarithm followed a normal distribution far better than did the untransformed version of this variable. For the other three variables (SAIFI with major events, SAIFI without major events, and SAIDI with major events), the statistical tests indicated that neither the transformed nor the untransformed variables conclusively followed a normal distribution. Third, finding no evidence that using the variables in their un-transformed state was superior to using them in their transformed state, the ability to provide an easy-to-explain interpretation of the regression coefficients led to decide to use the variables in their transformed state.

Visual Evidence that Transformed Variables Follow a Normal Distribution Better than Untransformed Variables.

From the standpoint of the regression analysis we sought to conduct, it is desirable that the dependent variables used in the analysis follow a normal distribution. Figures B-1 and B-2 show results from a graphical analysis that compares the observed data (i.e., the histogram bins) with theoretical normal (and log-normal) distributions (i.e., the curves shaded in blue). By visual inspection, we find that all four annual reliability metrics are more accurately represented by a log-normal distribution than by a normal distribution.

Statistical Tests for the Normality of the Distributions of Transformed and Untransformed Variables.

Tables B-1 and B-2 report results from three statistical testing methods—(1) KolmogorovSmirnov, (2) Cramer-von Mises, and (3) Anderson-Darling—commonly used to evaluate the assumed shape of a distribution. Table 1 shows that all tests conducted for all of the reliability metrics rejected the null hypothesis of normality with a high degree of confidence. Table B-2 shows that the null hypothesis of log-normality was rejected for SAIFI (with and without major events included) and SAIDI (with major events), but we fail to reject the null hypothesis for two of the three tests of SAIDI (without major events). To summarize, formal statistical testing indicated that SAIDI (without major events included) was best fit with a log-normal distribution. However—with the exception of SAIDI (without major events)—statistical testing rejected the null hypothesis of both normality and log-normality at a 99% or greater confidence level.

Table B- 1. Statistical Tests for Normality

	Reject Null Hypothesis of Normality?		
Reliability Metric	Kolmogorov -Smirnov	Cramer-von Mises	Anderson- Darling
SAIDI (w/o major events)	Yes***	Yes***	Yes***
SAIDI (with major events)	Yes***	Yes***	Yes***
SAIFI (w/o major events)	Yes***	Yes***	Yes***
SAIFI (with major events)	Yes***	Yes***	Yes***

Note: *** Rejects the null hypothesis at the .01 significance level.

Table B- 2. Statistical Tests for Log-normality

	Reject Null Hypothesis of Log-normality?		
Reliability Metric	Kolmogorov -Smirnov	Cramer-von Mises	Anderson- Darling
SAIDI (w/o major events)	No	No	Yes**
SAIDI (with major events)	Yes***	Yes***	Yes***
SAIFI (w/o major events)	Yes***	Yes***	Yes***
SAIFI (with major events)	Yes***	Yes***	Yes***

Note: *** Rejects the null hypothesis at the 0.01 significance level; ** Rejects the null hypothesis at the 0.05 significance level.

The Easy-to-Explain Interpretation of Regression Coefficients when Variables are Transformed.

The visual and statistical tests reported above indicate limited support in favor of using transformed variables. Importantly, they offer no support for the superiority of using untransformed variables.

Transformation of reliability metrics into logarithmic format allows for a natural interpretation of the estimated coefficients from the regression equations as percentages. For example, if an estimated coefficient for an explanatory variable has a value of 0.02, the interpretation is that a step change in that variable is correlated to a 2% increase in the reliability metric.

APPENDIX C. EXAMINATION OF OUTLIERS

We conducted a three-step analysis of the regression results to identify extreme and influential outliers, research circumstances that might have affected the reporting (values) of SAIFI and SAIDI, and tested the effect of removing outliers on the model results.

First, we flagged any observations that had a studentized residual that exceeded a pre-defined threshold of plus or minus three (UCLA 2011). In parallel, we carried out a Cook's statistical test to evaluate the size of the residual and influence (leverage) of the individual observations on the model results. Any observation with a Cook's D statistic greater than four divided by the sample size (n) was flagged as an influential and extreme outlier (UCLA 2011). We compiled a list of outliers that were flagged by *both* the studentized and Cook's statistical method simultaneously and carried out a deeper analysis on these observations.

Next, we reviewed the reliability event information to determine if any errors in information transcription occurred when we were collecting the data. No errors were found in the process of entering the data into our database from the source of the reliability event data.

We then looked into possible explanations for the extreme values to help us understand what was happening during these specific years and at these utilities. Table C-1 summarizes the number of extreme and influential outliers that were identified in the preceding steps. As shown in the table below, most of the outliers are explained by either a rare weather occurrence or by the characteristics of the utility service territory with these events leading to lower or higher values of the metrics when compared to other utilities.

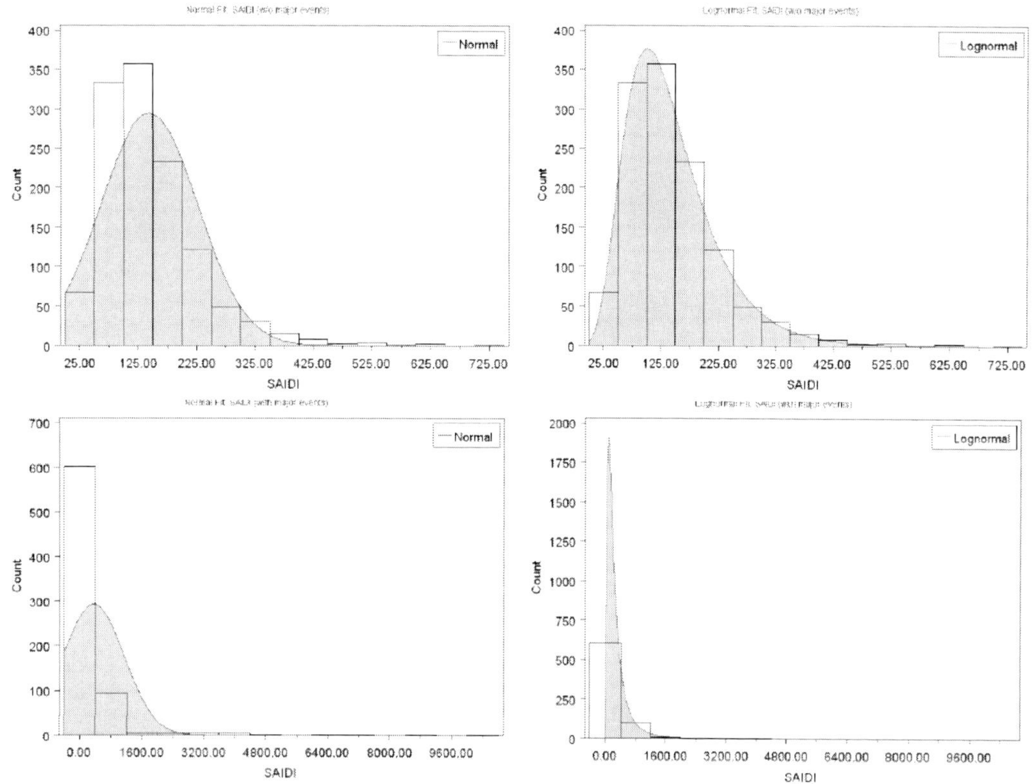

Figure B- 1. Graphical analysis of SAIDI with and without major events included.

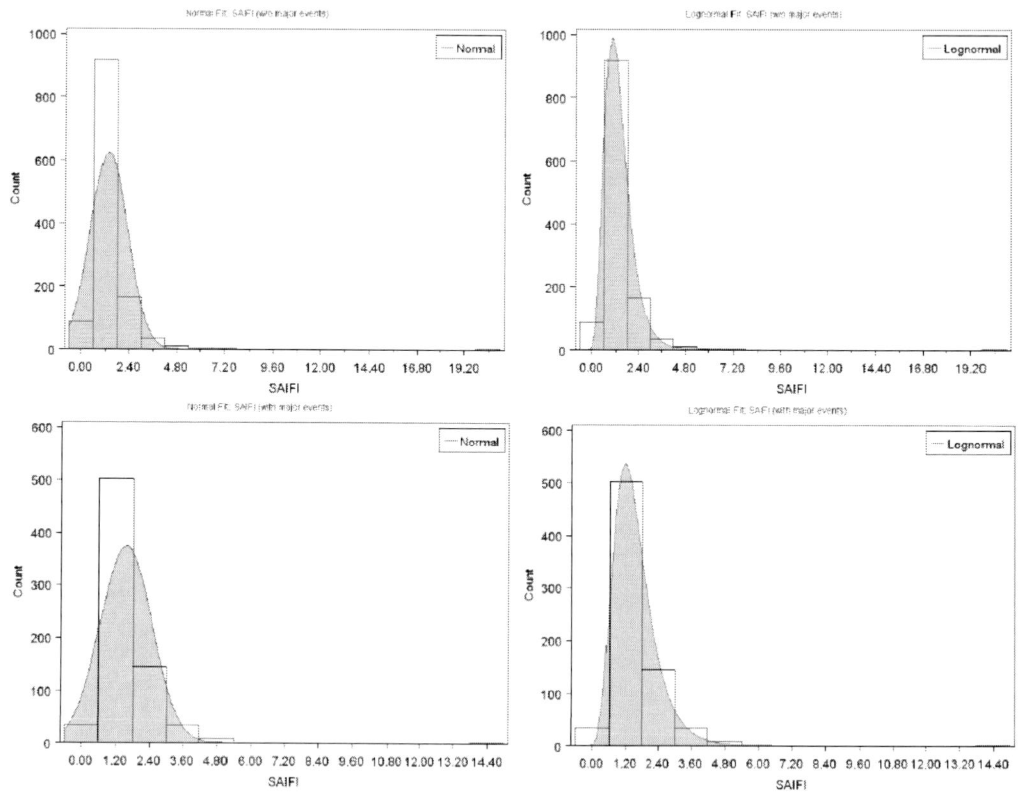

Figure B- 2. Graphical analysis of SAIFI with and without major events included.

Table C- 1. Summary Explanation of Identified Outliers

Dataset	Number of Extreme and Influential Outliers	Number of Utilities	Explanation
SAIDI with major events included	17	12	• 9 outliers due to severe storms, including two hurricanes • 1 outlier due to increased use of troubleshooting personnel that impacted the reliability metrics • No information on the remaining seven outliers
SAIDI without major events included	17	5	• 14 outliers attributed to characteristics of the service territory including a small territory size and increased use of troubleshooting personnel that impacted the reliability metrics • No information on the remaining three outliers
SAIFI with major events included	10	4	• Seven are from a single utility that attributes their anomalous metric values to the high concentration of distribution networks and large customer base • One was due to a large wind and snow storm • No information on the remaining two outliers
SAIFI without major events included	27	8	• 20 outliers attributed to characteristics of the service territory including things like a small territory size, high concentration of distribution networks, and representation of a large number of customers • No information on the remaining 7 outliers

Finally, we ran the regressions again using two methods to exclude outliers: 1) without the lowest and highest 1% of SAIDI (SAIDI) values (i.e., the 1%/99% exclusion method) and 2) without the extreme and influential outliers we identified in the Studentized and Cook's statistical analysis discussed. We found that the regression results did not significantly change when the outliers were removed according to these two methods. Table C-2 is a summary of the effects on the regression results after removing outliers using two different methods.

Table C- 2. Excluding Outliers and their Effect on the Pooled Regression Results

Pooled Regression (i.e., No Utility or Time Effects)	Outlier Exclusion Method	Effect of Excluding Outliers on Regression Results
SAIDI without major events included	1%/99% Method	R2 slightly increased from 0.04 to 0.05; No sign changes on regressors; No regressors lost or gained significance at the 10% level.
SAIDI without major events included	Cook's and Studentized Residual Tests	R2 slightly increased from 0.04 to 0.06; No sign changes on regressors; No regressors lost or gained significance at the 10% level.
SAIFI without major events included	1%/99% Method	R2 slightly increased.; After outlier exclusion, heating degree-days (HDD) became significant at the 10% level and year became marginally insignificant at the 10% level (p=0.109); Post OMS regressor was marginally significant at 10% level before excluding the outliers (p=.106), but definitely not significant after the outliers were removed (p=0.43).
SAIFI without major events included	Cook's and Studentized Residual Tests	R2 slightly increased from 0.07 to 0.09. No sign changes on regressors. After outliers were removed, heating degree-days (HDD) became significant at the 10% level. No other regressors lost or gained significance at the 10% level.

As a result of these three steps, we concluded that the outliers identified in our statistical analysis are valid observations.

We also determined that their removal did not significantly affect the pooled regression results. For these reasons, we chose not to remove any of these outliers from the statistical regression analysis presented in the main body of the report.

APPENDIX D. DETAILED RESULTS FROM REGRESSION ANALYSIS

Table D-1. One-way Fixed Effects Regression for the Effect of Sales, HDD, CDD, time, and OMS on Frequency and Duration of Grid Disruptions

	With Major Events			Without Major Events			
	I ln SAIFI		II ln SAIDI	III ln SAIFI		IV ln SAIDI	
Intercept	-61.9241 ***		-123.951***	-45.221	***	-46.4962	***
	(19.3904)		*(38.0343)*	*(10.2925)*		*(11.5011)*	
Sales	0.00295		0.006712	0.001054		0.002057	
	(0.0044)		*(0.0087)*	*(0.0036)*		*(0.0041)*	
HDD	2.109E-06		0.000028	-0.00002		-0.00004	
	(0.0001)		*(0.0001)*	*(0.0000)*		*(0.0000)*	
CDD	5.867E-06		-0.00005	0.000182	*	0.000236	**
	(0.0001)		*(0.0003)*	*(0.0001)*		*(0.0001)*	
YR	0.030683 ***		0.063937***	0.022749	***	0.025635	***
	(0.0097)		*(0.0190)*	*(0.0051)*		*(0.0057)*	
OMS	0.055153		0.381655***	-0.02486		0.173814	***
	(0.0636)		*(0.1248)*	*(0.0464)*		*(0.0517)*	
POST OMS	-0.02157		-0.03947	-0.0128		-0.01033	
	(0.0133)		*(0.0262)*	*(0.0092)*		*(0.0103)*	
Utility Effects	Yes		Yes	Yes		Yes	
R2	0.59		0.47	0.71		0.69	

Notes: Standard errors in parentheses. * p<0.10, ** p<0.05, *** p<0.01.

Table D- 2. Two-way Fixed Effects Regression for the Effect of Sales, HDD, CDD, and OMS on Frequency and Duration of Grid Disruptions

	With Major Events				Without Major Events			
	I ln SAIFI		II ln SAIDI		III ln SAIFI		IV ln SAIDI	
Intercept Sales	-0.41598		4.50384	***	0.542495		4.874969	***
	(0.3332)		*(0.6499)*		*(0.3664)*		*(0.4091)*	
	0.003169		0.007662		0.001042		0.001522	
	(0.0044)		*(0.0085)*		*(0.0037)*		*(0.0041)*	
HDD	0.000041		-5.51E-06		-0.00002		-0.00003	
	(0.0001)		*(0.0001)*		*(0.0001)*		*(0.0001)*	
CDD	-0.00016		-0.00035		0.000051		0.000215	*

Table D- 2. (Continued)

	With Major Events				Without Major Events			
	I ln SAIFI		II ln SAIDI		III ln SAIFI		IV ln SAIDI	
	(0.0002)		*(0.0003)*		*(0.0001)*		*(0.0001)*	
POST OMS	-0.01241		-0.02826		-0.00842		-0.00889	
	(0.0133)		*(0.0260)*		*(0.0093)*		*(0.0104)*	
Year 1	-0.21793	**	-0.33234	*	-0.16384	***	-0.13058	**
	(0.1028)		*(0.2005)*		*(0.0585)*		*(0.0654)*	
Year 2	-0.08249		-0.29269		-0.13054	**	-0.15683	**
	(0.0980)		*(0.1912)*		*(0.0573)*		*(0.0640)*	
Year 3	0.050256		0.050184		-0.04065		-0.11338	*
	(0.0897)		*(0.1750)*		*(0.0556)*		*(0.0621)*	
Year 4	0.085979		0.130981		-0.01804		-0.08814	
	(0.0841)		*(0.1640)*		*(0.0514)*		*(0.0573)*	
Year 5	0.038003		-0.06511		-0.03443		-0.01921	
	(0.0804)		*(0.1569)*		*(0.0496)*		*(0.0552)*	
Year 6	0.093184		0.027743		0.045398		0.018774	
	(0.0801)		*(0.1562)*		*(0.0519)*		*(0.0580)*	
Year 7	0.193096	**	0.296795	*	0.062783		0.039727	
	(0.0835)		*(0.1629)*		*(0.0559)*		*(0.0625)*	
Year 8	0.133398	*	0.215998		0.025579		-0.01463	
	(0.0715)		*(0.1394)*		*(0.0476)*		*(0.0531)*	
Year 9	0.232577	***	0.572835	***	0.100408	**	0.138323	***
	(0.0648)		*(0.1264)*		*(0.0436)*		*(0.0488)*	
Utility Effects	Yes		Yes		Yes		Yes	
R2	0.61		0.50		0.71		0.70	

Notes: Standard errors in parentheses.*p<0.10, ** p<0.05,***p<0.01.

REFERENCES

Amin, M. 2008. "Challenges in Reliability, Security, Efficiency, and Resilience of Energy Infrastructure: Toward Smart Self-Healing Electric Power Grid." *Power and Energy Society General Meeting - Conversion and Delivery of Electrical Energy in the 21st Century*. pp.1-5, July 20-24. http://ieeexplore.ieee.org/stamp/stamp.jsp?tp=&arnumber=4596791&isnumber=4595968.

Baltagi, B. H. and Y. Chang. 1994. "Incomplete Panels: A Comparative Study of Alternative Estimators for the Unbalanced One-Way Error Component Regression Model." *Journal of Econometrics,* 62(2): 67-89.

Buse, A. 1973. "Goodness of Fit in Generalized Least Squares Estimation," American Statistician, 27. pp. 106-108.

Cameron, A. C., and P. Trivedi. 2009. Microeconometrics Using Stata. TX: Stata Press.

Energy Information Administration (EIA). 2010. "Form EIA-861 Final Data File for 2009." DOE/EIA. http://ei-01.eia.doe.gov/cneaf/electricity/page/eia861.html.

Environmental Protection Agency (EPA) 2011. "Emissions & Generation Resource Integrated Database (eGRID)". http://www.epa.gov/cleanenergy/energy-resources/egrid/index.html.

Eto, J. E. and K. H. LaCommare. 2008. *Tracking the Reliability of the U.S. Electric Power System: An Assessment of Publicly Available Information Reported to State Public Utility Commissions*. Berkeley CA: Lawrence Berkeley National Laboratory Report LBNL-1092E. October. http://certs.lbl.gov/certs-rtina-pubs.html.

Fisher, E., J. Eto, K. Hamachi-LaCommare. 2012. "Understanding Bulk Power Reliability: The Importance of Good Data and A Critical Review of Existing Sources." *Proceedings of the 45th Annual Hawaii International Conference on System Sciences*. Wailea, HI. Jan. 1-4, 2012.

Greene, W. 2000. Econometric Analysis (Fourth Edition). Upper Saddle River NJ: Prentice-Hall.

Hausman, J. A. 1978. "Specification Tests in Econometrics," Econometrica, 46: 1251-1271.

Hines, P., J. Apt, and S. Talukdar. 2009. "Large Blackouts in North America: Historical Trends and Policy Implications." *Energy Policy*, v. 37, pp. 5,249-5,259.

IEEE Power Engineering Society. 2004. *IEEE Std 1366-2003 IEEE Guide for Electric Power Distribution Reliability Indices.* ISBN 0-7381-3890-8 SS95193. New York: Institute of Electric and Electronics Engineers, Inc. May 14. 35 pages.

LET Systems. 2006. "Requirements for the Implementation of an Outage Management System (OMS) Whitepaper." January. http://www.letsys.com/img/oms implementation requirements white paper.pdf.

McGranaghan, M., A. Maitra, C. Perry, A. Gaikwad. 2006. "Effect of Outage Management System Implementation on Reliability Indices." *Transmission and Distribution Conference and Exhibition, 2005/2006*

IEEE PES. pp.1,208-1,211, May 21-24. http://ieeexplore.ieee.org/stamp/stamp.jsp?tp=&arnumber=1668677&isnumber=34941

National Climatic Data Center (NCDC). 2011. U.S. Department of Commerce/NOAA: U.S. Climate Normals; Area-Weighted State, Regional, and National Temperature. Accessed May 2011 at http://cdo.ncdc.noaa.gov/cgi-bin/climatenormals/climatenormals.pl

SAS. 2011a. Specification Tests. Accessed June 27 at: http://support.sas.com/documentation/cdl/en/etsug/63348/HTML/default/viewer.htm#etsug_pane l sect039.htm.

SAS. 2011b. The One-way Random Effects Model. Accessed June 21 at: http://support.sas.com/documentation/cdl/en/etsug/60372/HTML/default/viewer.htm#etsug_pane l sect031.htm.

UCLA. 2011. Introduction to SAS. UCLA: Academic Technology Services, Statistical Consulting Group. Accessed September 2011at: http://www.ats.ucla.edu/stat/sas/notes2/.

Ventyx. 2011. Velocity Suite Database System, Boulder. Accessed May 23.

Wansbeek, T., and A. Kapteyn. 1989. "Estimation of the Error-Components Model with Incomplete Panels," Journal of Econometrics, 41, 341-361.

Wooldridge, J. 2002. Econometric Analysis of Cross Section and Panel Data, MIT Press.

End Notes

[1] Others observe that the data on bulk power system reliability relied on by studies such as these are sometimes inconsistent, incomplete and inaccurate (Fisher, et al. 2012).

[2] In some instances when SAIDI was not reported, the Customer Average Interruption Duration Index (CAIDI) was collected to derive SAIDI using the simple mathematical expression CAIDI = SAIDI/SAIFI.

[3] Temperature records came from observation stations located in climatologically homogenous regions within a state. The station's observations are weighted by the area of its climate region as a proportion of the state's area. This produces a weighted average for temperature in the state. For further details on the weighting procedures, see NOAA National Climatic Data Center (2011).

[4] The electricity sales information from the EIA 861 form is also housed in a large database supported by Ventyx (Ventyx 2011).

[5] Our sample is influenced by the decision to collect reliability information reported to state regulatory utility commissions because commission rules usually specify how data are to be reported and, in particular, whether the IEEE Standard 1366-2003 or another set of reliability data definitions will be used. Many utilities rely on IEEE Standard 1366-2003 for internal uses of reliability metrics and at the same time report reliability data to their state regulatory utility commissions using different definitions, as required by commission rules.

[6] In every instance in which a utility relied on IEEE Standard 1366-2003 to report its reliability metrics, the standard was used to prepare the metrics for each year for which data were obtained. In many instances, this meant that the utility had recalculated its reliability metrics using the standards for the years prior to the utility's decision to use the standard.

[7] Figure 3 in Section 2 shows the number of utilities for which we had 10 years of reported reliability data for each of the four reliability metrics.

[8] Appendix A compares a variant of this analysis of customer-weighted means, which enables a direct comparison to a closely related analysis conducted by the IEEE Distribution Reliability Working Group.

[9] In order to understand the effect on reported reliability using IEEE Standard 1366-2003, utility effects cannot be included in the model because utility effects take account for all systematic differences among utilities, including whether or not the utility used IEEE Standard 1366-2003.

[10] While reasonable for the purposes of this chapter, we plan to explore the assumption of linearity in future research.

[11] Note that because this is an unbalanced data set, the Breusch-Pagan test for random effects is not appropriate here (SAS 2011a).

[12] See Appendix D for fixed effect results; as expected, the coefficient estimates are similar but less efficient (that is, they are not statistically significant).

[13] Model 1 restricts the time trend to be linear. In the appendix, we present results from a model that includes year fixed effects rather than a linear time trend. The estimates of the time fixed effects increase in a relatively linear fashion, but the loss in degrees of freedom results in estimates that are not statistically significant. As noted in an earlier footnote, we plan to explore the assumption of a linear time trend in future research.

[14] A high-level analysis of identified outliers was also performed to assess the impact on the regression results and to understand the circumstances behind the outlier. Please see Appendix C.

[15] We plan to explore other means for measuring a learning effect, such as consideration of additional years following installation or upgrade of an OMS.

In: Electricity Reliability in the United States
Editor: Emmanuel Martinez
ISBN: 978-1-62417-149-9

Chapter 2

UNDERSTANDING BULK POWER RELIABILITY: THE IMPORTANCE OF GOOD DATA AND A CRITICAL REVIEW OF EXISTING SOURCES*

Emily Fisher, Joseph H. Eto and Kristina Hamachi LaCommare

ABSTRACT

Bulk power system reliability is of critical importance to the electricity sector. Complete and accurate information on events affecting the bulk power system is essential for assessing trends and efforts to maintain or improve reliability. Yet, current sources of this information were not designed with these uses in mind. They were designed, instead, to support real-time emergency notification to industry and government first-responders. This chapter reviews information currently collected by both industry and government sources for this purpose and assesses factors that might affect their usefulness in supporting the academic literature that has relied upon them to draw conclusions about the reliability of the US electric power system.

---

* This is an edited, reformatted and augmented version of a Lawrence Berkeley National Laboratory publication (LBNL-5125E) dated October 2011.

1. INTRODUCTION

The reliability of the bulk power system is of critical importance to the electricity system in the U.S. because it relies on high-voltage power transportation. Understanding the reliability of the bulk power system, how to measure it, and how that reliability is changing over time, is important for developing and assessing technologies and policies that seek to affect reliability.

Meaningful assessment of reliability depends on comprehensive and accurate data. Currently, there are two major data collection efforts that focus on the reliability of the bulk power system in the U.S. The Department of Energy (DOE) collects data through the Electric Emergency Incident and Disturbance Report (via Form OE-417). The North American Electric Reliability Corporation (NERC) collects data through the reporting requirements contained in the Disturbance Reporting reliability standard, EOP-004, which is known as the Interconnection Reliability Operating Limit and Preliminary Disturbance Report.

Many authors have used these data to assess characteristics of and longitudinal trends in bulk power system reliability. By and large, these authors take the completeness and accuracy of these data as given. Yet both sources of data were developed originally to support emergency notification. It is not evident from their design and implementation that they were intended to serve as comprehensive repositories of definitive information on bulk power system reliability. Closer examination of these data for completeness and accuracy, therefore, is warranted to help us understand what kinds of analysis can be performed using these data, and evaluate the claims made by authors who use the data to support their analysis. Examining current reliability datasets can also indicate potential changes in future data collections that may better support reliability analysis.

The objective of this chapter is to begin exploring what can and cannot be learned from analyzing the two current sources of data on the reliability of the bulk power system. To do this, we first review the history, purpose, and current reporting requirements for the DOE and NERC reliability data. In the third section, we examine the completeness and accuracy of DOE and NERC data, first by comparing them to one other, and second by comparing the NERC data to more detailed investigations that were subsequently conducted by NERC on a handful of events. We then, in the fourth section, review the literature that has relied on these data and use our findings to draw preliminary conclusions regarding the claims that have been made in the literature. In the fifth section, we outline possible directions for improving the comprehensiveness and accuracy of data on the reliability of the bulk power system.

2. CURRENT DOE AND NERC BULK POWER SYSTEM RELIABILITY EMERGENCY REPORTING

The DOE began collecting data on the reliability of the bulk power system through the EIA-417 form in the late 1970's. At that time the Energy Information Administration (EIA) administered the form. When the DOE's Office of Electricity Delivery and Energy Reliability (OE) was established in 2003, the office, among other things, took over collection and validation of this bulk power system reliability data and renamed the form OE-417.

The purpose of the form is "to meet [DOE's] overall national security and Federal Emergency Management Agency's National Response Framework responsibilities," by "obtain[ing] current information regarding emergency situations on U.S. electric energy supply systems." [1] Another way to define the form's purpose is to alert the Federal government when the bulk power system has transitioned from a normal state of operation to an emergency state of operation. The information submitted on the form is circulated to various government agencies and to the public. For example, the data are used in the OE Spot Report, which is circulated only to other government agencies, and the OE Situational Report, which is publically available. Annual data is also available publically on the OE website.[1]

Entities that are required to submit OE-417 include reliability coordinators, balancing authorities, utilities and some generating entities. An initial report is required either one or six hours following a triggering event (described below as reporting criteria) and a final report is due 48 hours following the triggering event.[2] OE staff take a variety of measures to improve the quality and completeness of the reporting, including contacting respondents to quality-check data and utilities that staff learn have experienced an event but have not filed an OE-417 form. According to OE staff, data posted on the OE website are the best available, developed from the initial and final reports as well as the follow-up work by OE staff. However, staff also note that there are limited resources available for enforcement of the reporting requirements and do not know how many events go unreported.

Historical changes in the requirements and procedures for filing have produced discontinuous jumps in the number of events reported. For instance, in the 1990's municipal utilities and cooperatives were added to the required respondents, resulting in an increase in respondents from around 50 to around 200.[3] According to OE staff, when NERC instituted mandatory reliability reporting in 2007, there was a large increase in reported events to DOE.[4]

Reporting criteria (i.e., triggering events) for filing form OE-417 fall into four main areas: intentional attacks on the system, system failures (e.g., blackout) or voltage events, load loss and/or emergency reduction, and fuel supply emergencies. For the more severe or serious events—actual physical and cyber attacks resulting in interruptions, impacts to critical infrastructure or operations; complete operational failure; separation of the system (i.e., islanding); uncontrolled firm load interruption of 300 MW or more; emergency load shedding of 100 MW or more; system-wide voltage reductions; and public appeals to reduce electricity use—initial reporting is due within one hour of the start of the event. For the less severe events, including suspected attacks, loss of electric service to 50,000 customers for one hour or more, and fuel supply emergencies, initial reporting is due within six hours of the start of the event. Any event reported on form OE-417 to DOE must also be submitted to NERC, under the NERC Disturbance Reporting reliability requirement, EOP-004.

NERC began collecting reliability data in 1984 through System Disturbance Reports. In 2007, submission of bulk power events became mandatory through the Disturbance Reporting requirement, EOP-004. [2] The NERC EOP-004 is one of many reporting and reliability requirements that were developed by NERC in their capacity as the Electric Reliability Organization.

The purpose of NERC EOP-004 is to make reliability event data available for study and analysis in order to avoid similar problems in the future. More types of entities are subject to the reporting requirements of EOP-004 than are required to submit form OE-417. In addition to the reliability coordinators, balancing authorities, utilities (load serving entities), and some generators that are required to submit form OE-417, NERC's reporting requirements also apply to transmission operators, all generator owners, and regional reliability organizations.

EOP-004 is mandatory and requires that all events reported to DOE under OE-417 must also be reported to NERC. In addition to the DOE criteria for reporting the following types of events also trigger mandatory EOP-004 reporting: loss of generation (2,000 MW or more in the Eastern or Western

Interconnection, 1,000 MW or more in ERCOT); loss of other components that result in loss of 300 MW of firm load (or a lesser amount for small entities), modification of operating procedure, under-frequency or under-voltage situations, or other serious outcomes; firm load shedding of 100 MW or more to maintain continuity of the bulk power system; system islanding; and any action resulting in voltage excursions, component damage, or interference with automated system protection. For most of these events a preliminary report must be submitted to NERC within 24 hours. A final report is due 60 days after the event. Event data reported under EOP-004 are available in annual files on the NERC website, beginning with 1992 data.[5] Year 2008 and 2010 data are being reviewed and processed by NERC, and, as of August 2011, have not been released publically.

OE-417 and the NERC Disturbance Report Form request nearly identical information, including a description of the triggering event, date and time of event and of restoration, whether the event originated in the reporting entity's system, and amount of load and number of customers interrupted. In fact, since 2007, DOE has allowed submission of the NERC Disturbance Report Form in lieu of OE-417. NERC requires submission of form OE-417 as part of the reporting requirements of standard EOP-004.[6]

There are, however, some slight differences in the reports. For instance the NERC Disturbance Report Form asks for system frequency just prior and following the event, and OE-417 does not. OE-417 asks for high voltage substations lost during the event, and the NERC Disturbance Report Form does not.

By design, the intent of these forms is to collect information on reliability events, but not necessarily the exact same events. At the same time, these datasets may not be complete or accurate, in the sense that events that should have been reported to DOE or NERC may not have been, or events that were reported may contain information that does not exactly reflect what occurred during the event. Researchers need to understand what these datasets contain and how complete and accurate they are before using them in reliability analysis. In the following section, we begin this examination.

3. Analysis of NERC and DOE Reliability Data

On the face of it, various factors suggest neither the DOE nor the NERC dataset presents a complete or fully accurate picture of bulk power system reliability. From the standpoint of completeness, data collection by NERC has not always been mandatory. Requirements for who must file the OE-417 have changed over time. While OE-417 reporting is mandatory, and DOE staff takes measures to contact utilities that may not have filed a report when required, DOE does not have the resources to enforce compliance strictly. NERC Disturbance Report data from 2008 have not been released publically, and (at the time this chapter was written in the Summer of 2011) no anticipated release date is set.

From the standpoint of accuracy, the principal factor is the short reporting time frame (although efforts are made to validate information and release publically only what is considered to be the most accurate). Considering, for instance, the number of customers that experience an interruption, in the absence of automated outage management systems (which may or may not be the sources of information reported on the forms), or without an in-depth post-event investigation by the respondent, inaccuracy is possible.

In this section we present our analysis of bulk power system reliability data collected by NERC and DOE from 2000 to 2009 (excluding 2008 because NERC has not released data from that year). We focus only on events that meet DOE's criteria of 50,000 or more customers interrupted and/or 300 MW of firm load shedding. Since 2007, all events reported to DOE under OE-417 criteria were required to be submitted to NERC as well; before 2007, while loss of 50,000 customers was not a specific criterion for NERC reporting, an event of that magnitude would likely meet other NERC criteria.

First we compare the NERC and DOE data to one another and show evidence that the datasets are incomplete and potentially inaccurate. We do this in two ways: first by identifying how often events are only reported to one institution or when the same event has different information reported describing the event; and second, by using each dataset to calculate common reliability metrics. Then we compare NERC emergency disturbance reporting with more detailed analyses for several events and discuss the types of corrections that were made following these more detailed post-event investigations.

The goal of our analysis is to determine whether the datasets contain accurate and complete information. If the same event is reported with different load loss information, this is an indication that at least one of the datasets is not accurate. If an event is reported in one form but not another when it meets the criteria of both, it is an indication that the datasets are not complete. If the datasets are not complete, some reliability events are missing from the record (and from any subsequent analysis).

3.1. Comparison of Events Reported to NERC and DOE

To get a preliminary sense for the completeness and accuracy of the two datasets, we compare the reported events to determine if they were reported to both NERC and DOE, and whether the events were reported in the same way to both. First, we identify where NERC and DOE records referred to the same event, based on comparing event start date, and utility name or location. When there seems to be a match, we determine whether the event parameters (start date and time, restoration date and time, utility, number of customers interrupted and customer load lost) are consistent between the two reports. If all seven parameters match, the two event records are considered a "Match." Where at least one of these parameters is not the same, they are considered a "Match with differences." This includes differences that may have been a result of input error (either by the respondent or by DOE/NERC staff that entered the data into their respective databases). For instance, for one event the number of customers was reported as 159,019 to NERC and 157,019 to DOE. In a similar example, the number of customers interrupted was reported as 8,110 to NERC and 88,110 to DOE. In other cases, the event start times differ by 12- hours, which may be a result of incorrect interpretation of a 24-hour clock. Because it is difficult to determine whether there was an error in reporting or an intentional difference, these reports were considered "Match with differences."

For this analysis we exclude events that took place outside the continental US, because reporting requirements are not consistent.[7] In addition, we consider only events with 50,000 or more customers interrupted and/or loss of 300 MW or more of firm load. We apply this screening after we matched events, so in some cases an event reported to one institution does not meet the screening criteria but is a match to an event reported to the other institution that did. Our goal is to include as many reported events we could reasonably justify should be reported to both DOE and NERC. If an event meets the criteria as reported to one of the institutions it is reasonable to assume the event should be reported to other institution.

This variation in reporting of the same event to NERC and DOE suggests that in some cases where an event is reported to one institution and does not meet our screening criteria, it may not meet the screen simply because the event parameters were entered incorrectly. As a result our screening strategy may remove significant events that should have been reported to both NERC and DOE because the number of customers interrupted or loss of load was not reported. We took this conservative approach in order to not over-state the mismatch between the datasets; in light of this, the "No Match" values below may be higher than we present, suggesting the datasets may be less complete than appears in this analysis.

Figure 1 summarizes the matching of data between NERC and DOE reports. The number of loss of load events reported to both institutions grows over time, from one or two events in the early 2000's, to 22 in 2009. However, in each year there are some events reported to NERC that are not reported to DOE, even though they meet the DOE reporting criteria. Also, in 2007 and 2009 when all OE-417 reports are required to be submitted to NERC, some events are not. This suggests each dataset is missing some events, and therefore both are incomplete. A careful look at 2009 data reveals three events in the NERC dataset with more than 50,000 customers interrupted not found in the DOE dataset. Similarly, nine events reported to DOE were not found in the EOP-004 dataset; there should have been none, because as of 2007 all OE-417 events are required to be reported to NERC.

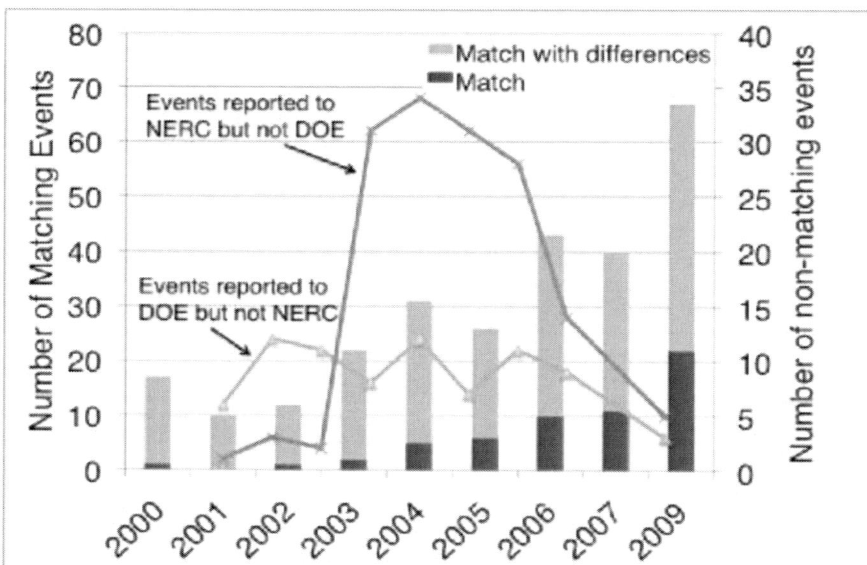

Figure 1. Matching of event reported to NERC and DOE.

This matching exercise also provides insight on whether the datasets are accurate. As Figure 1 illustrates, the number of events that fall under the "Match with differences" is not decreasing over time. This means there are still many reports of events in the DOE and NERC datasets that are inconsistent in one of the important parameters: number of customers interrupted, amount of load lost, or length of interruption. These are critical parameters for analyzing reliability, and if they are inconsistently reported, it means at least one of the datasets contains inaccuracies.

3.2. Comparison of Bulk Power System SAIFI and SAIDI Using NERC and DOE Reliability Data

One important measure of reliability is how often an end-use customer experiences power interruptions, and what the duration of those interruptions are. To determine whether the DOE and NERC reliability datasets are telling a consistent story about customer interruptions, we calculate standard metrics used to report distribution system reliability for each dataset and compare them. We do not seek to draw conclusions about the reliability of the bulk power system or how it is changing over time, but rather to examine whether it is appropriate to use the existing NERC or DOE data to examine customer interruptions.

System Average Interruption Frequency Index (SAIFI) and System Average Interruption Duration Index (SAIDI) are two common metrics of reliability of power systems. [3] SAIFI measures how many times - on average - an end-use customer experiences a power interruption over a given period of time. SAIDI measures how long the average customer is without power over a given period of time. The equations to calculate SAIFI and SAIDI are as follows:

$$\text{SAIFI} = \frac{\sum \text{Total Number of Customers Interrupted}}{\text{Total Number of Customers Served}}$$

$$\text{SAIDI} = \frac{\sum \text{Customer Interruption Durations}}{\text{Total Number of Customers Served}}$$

We calculate annual SAIFI by summing all events in the NERC and DOE datasets that have a value for number of customers interrupted. We obtain the total number of customers in the country for that year (taken from EIA form 861).

If NERC and DOE reliability data are consistent and accurate with respect to customer interruptions originating on the bulk power system, the annual SAIFI would be the same. However, as can be seen in Figure 2, they are different. There is as much as a 54 percent difference between the two SAIFI values in a given year. This difference is due to two factors: the two datasets contain at least some different events, and a different number of interrupted customers reported for matched events. The difference in SAIFI cannot be explained by a consistent bias in the datasets, e.g., one dataset being more "comprehensive" than the other, as suggested in the literature [4], because in some years SAIFI is higher for one dataset and in other years it is lower.

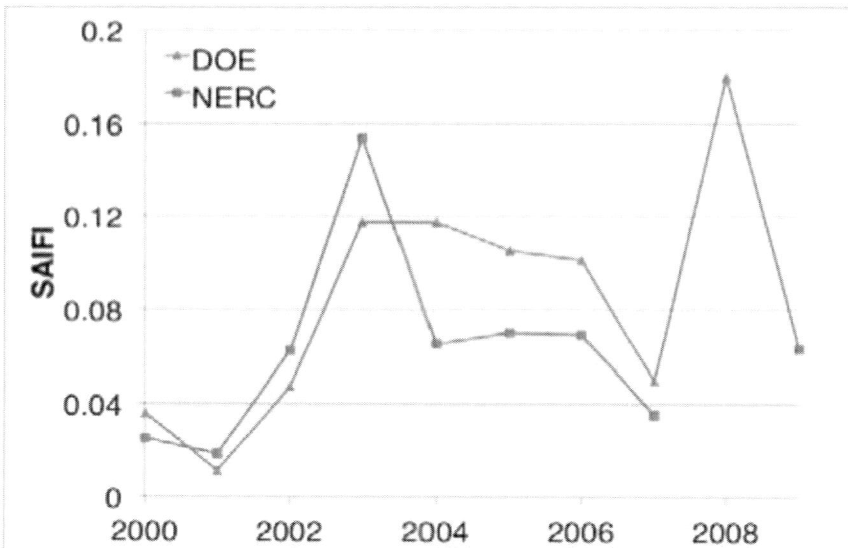

Figure 2. Annual SAIFI calculated from NERC and DOE bulk power system reliability data.

We calculate SAIDI with all events in the NERC and DOE datasets for which we can determine the event duration; in other words, all events with interruption start and end date and time. However, not all customers are without power for the entire time between the start and end of an event. As the utility responds to the event, power is restored sequentially to affected customers, until all customers are completely restored and the event is considered over. We estimate this function, customers restored as a function of event duration, in two ways: linear restoration and exponential restoration. Linear restoration assumes there is a constant rate at which service is restored to customers.

$$\text{Customer Mins Interrupted (linear)} = \frac{[\#\text{ Customers Affected}] * [\text{Mins of Interruption}]}{2}$$

Exponential restoration assumes customers are restored slowly at first, but at an increasing rate.

$$\text{Customer Mins Interrupted (exp)} = \frac{[\#\text{Customers Affected}]}{\{[-\ln(0.01)]/[\text{Mins of Interruption}]\}}$$

We calculated SAIDI using both methods, and the results are shown in Figure 3.

The linear model of customer interruption minutes always results in a higher calculated SAIDI than that calculated with the exponential model; thus these two models can be seen as reasonable bookends. These are not the only two methods for estimating how long each customer was without power, but they are adequate for our purposes. For this analysis we are interested in the comparison of SAIDIs calculated with the NERC and DOE datasets, not the actual values of SAIDI resulting from the calculated restoration time.

As with the SAIFI comparison, if NERC and DOE reliability data are complete and accurate, the annual SAIDIs should be the same for a given year and customer restoration methodology. However, as can be seen in Figure 3, they are not. This disparity arises because there is a different number of events in each dataset as well as variation in the number of reported customers interrupted and duration for the matching events. As with the SAIFI analysis, we do not observe a consistent bias in the data. In some years DOE data results in a higher SAIDI, in other years NERC does. The percent difference between the two SAIDI values is as high as 192 percent.

3.3. Selected Comparison of NERC Emergency and Power-Event Analysis Event Reports

From the standpoint of assessing trends in or impacts of technologies and policies on bulk power system reliability, one weakness of the reliability data collected by NERC and DOE is that they must be reported very quickly after an event begins. There is not enough time in the reporting schedule to allow for detailed analysis of the causes or extent of the event; as noted earlier, subsequent analysis of these topics is not the purpose these reports were intended to serve. Indeed, submitting a report containing hurriedly estimated values can be challenging enough when utility personnel are primarily, and justifiably, concerned with restoring power to customers.

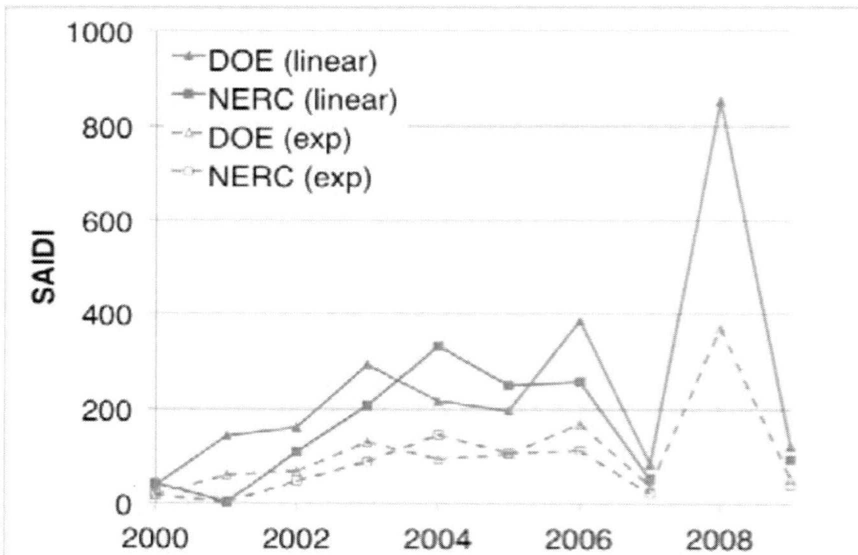

Figure 3. Annual SAIDI calculated from NERC and DOE bulk power system reliability data.

Conducting post-event analysis does not guarantee complete or more accurate values for event descriptors like number of customers interrupted or amount of load lost. However, if these values are revised as a result of post-event analysis, that must mean there was evidence that the original estimates could be improved.

To explore the kinds of changes that can be made to an event record when there is time to do a detailed analysis, we compare emergency reports with detailed analysis performed by NERC's Disturbance Analysis Working Group for 13 events from 2000 to 2002 (the most recent year these detailed analyses are available).[8]

We compare the following parameters from the two reports: number of customers, time event began, and time of restoration. Not every event has data in each category.

Table 1. NERC emergency and detailed disturbance reports: number of customers

Event	Date	# Customers	
		System Disturbance Report	DAWG Event Analysis
Public Service Co. NM Fire	3/18/00	600,000	660,000
TransEnergie short circuit	1/17/01	234,000	284,671
Saskatchewan Power Co. generation separation	3/10/01	246,000	248,588
ConEd NY - Terrorist Attack	9/11/01	12,000	13,000
WECCAZNMSNV contractor accident	8/2/02	350,000	348,360
WECCNWPP severe weather	12/26/02	Not reported	140,000

The number of customers interrupted is reported in five emergency reports and seven detailed reports. The difference in these values range from 1,640 (a near-zero percent change) to 50,671 (a 22 percent change). Number of customers lost is an input to both the SAIDI and SAIFI calculations, so inaccuracies in this number can affect calculation of these metrics.

Table 2. NERC emergency and detailed disturbance reports: time start

Event	Date	Time-Start	
		System Disturbance Report	DAWG Event Analysis
Public Service Co. NM Fire	3/18/00	15:57	15:50
VA Power Co. Substation Fire	4/1/00	16:47	16:46
NE WI, Up MI Peninsula Islanding	11/14/01	6:35	6:36
Jacksonville Electric Authority equipment failure	4/29/02	16:21	15:50

Table 3. NERC emergency and detailed disturbance reports: time restored

Event	Date	Time-Restored	
		System Disturbance Report	DAWG Event Analysis
New Brunswick Power Salt and Rain	12/20/00	17:28	16:59
TransEnergie short circuit	1/17/01	20:13	19:44
WECC line fault	8/1/01		6:44
Jacksonville Electric Authority equipment failure	4/29/02	4/29/02 23:59	4/30/02 1:30
MAIN - gen loss	7/26/02	19:00	16:33
WECCAZNMSNV contractor accident	8/2/02	Not reported	9:54
WECCNWPP severe weather	12/26/02	Not reported	15:13

Event start time is reported in emergency and detailed reports for four of the events. Of these, the difference in start time ranged from 1 minute to 31 minutes. Restoration time is reported in four emergency reports and seven detailed reports. The differences in restoration times range from 29 minutes to 147 minutes. Revision of event duration between the emergency and detailed reports can be calculated for six events. These revisions range from a change of one minute (a 0.3 percent change in duration) to 147 minutes (67 percent difference in duration). Interruption duration is an input to SAIDI, so a revision in this value would change that metric.

The differences identified here between events reported on emergency reports versus a more detailed analysis are significant in some cases. For all events we looked at, the values for customers interrupted, start time and/or restoration time were revised, at least to some degree, upon more detailed investigation. This indicates that emergency reporting, while important for emergency response, may not be appropriate as an archival record.

3.4. Summary of Analysis

This section presented our analysis of the NERC and DOE datasets and found they are not complete nor fully accurate from an archival perspective.

To some degree, the NERC and DOE reliability data are intentionally not complete: they include events that meet the applicable reporting criteria and exclude events that do not. However, as our comparison in section 3.1 indicates, events that should have been reported to both institutions in some cases were not. Over the time period we examine, the number of events missing from the DOE data are as high as 34 (in 2004); events missing from NERC data are as high as 12 (in 2001 and 2004). In section 3.2, we find that reliability metrics derived from the two datasets are inconsistent, in part because each contains at least some distinct reliability events not found in the other. In addition, policy shifts (e.g., expanding the pool of required respondents) and anecdotal evidence (e.g., noted increase in DOE submissions after NERC reporting was made mandatory) suggest the datasets do not contain a complete record of reliability events. Additionally, NERC has not released data from some periods of time, including 2008 and 2010.

Our analysis also finds the NERC and DOE reliability data are not fully accurate either. In many cases, reports of the same event submitted to NERC and DOE contain different number of customers interrupted, start and restoration times. We also find events where these parameters filed on emergency reports are revised after more detailed analysis. Because these event parameters are critical for analyzing reliability (in particular for calculating customer interruption metrics) using data that are not accurate may produce unreliable or misleading results.

However, there are many examples in the literature of reliability analysis conducted with these data. In the following section we examine this literature and assess the possible impact of incomplete or inaccurate source data may have on the findings.

4. Implications of Our Findings for the Literature

Many researchers use NERC reliability data, sometimes in concert with DOE data, to analyze the reliability of the bulk power system. We review their findings by organizing them under two categories: ones that conduct straightforward time-series or simple statistical analysis as evidence of bulk power system reliability, and ones that examine the nature of interruptions in the system in order to identify an explanatory or descriptive model.

In the first category, several publications have used NERC and DOE reliability data as evidence that the bulk power system is becoming less reliable. [4- 6] Many point to the increase from 1991 to 2005 in the *reported* number of significant interruptions and interruptions affecting more than 50,000 customers as evidence. [7] claims that patterns in NERC reliability data is evidence NERC's new reliability requirements have not improved reliability "yet." [8] shows NERC data (1984-2000) in cumulative probability of transmission and generation-related failures as motivation for their paper, encouraging focus on continuation of services in a blackout rather than only aiming to reduce or eliminate blackouts. [9] constructs time series-based statistical analysis with NERC data from 1990-2004 as input to risk management policies. The authors regress against time and customers to identify what factors affect the length of an interruption and how many customers are affected.

[10] conducts a statistical analysis of NERC interruption data. The authors make several claims: the frequency of large blackouts has not decreased between 1984 and 2006, there may be correlation between interruptions and season and time of day, there has been an increase in peak-hour and peak-season interruptions, there is evidence of a power-law relationship between interruption size and frequency (in support of SOC), and there is no correlation between interruption size and duration. The authors explain how they "scrub" the data, by focusing on only large events; controlling for demand growth, supply shortages and extreme natural events; and extrapolating incomplete data entries. Even still they make the assumption that, "all of the significant cascading failures are included in the NERC records, as required by regulation." [10]

Our findings from section 3—that the source data are incomplete and inaccurate—have the most potential ramifications for the time-series analysis since the consistency of reporting over time is fundamental to that kind of effort. Any finding based on a time-series analysis of the NERC or DOE reliability data is likely based on incomplete data, which may embed a bias in the analysis. The frequency of large blackouts could be quite different than what is evident in the data, if events have been underreported. Also, and more importantly, if the rate of reporting has increased (because of policy or other changes), a seeming increase in interruptions over time may be the result of improved reporting, not degrading reliability. Several papers conclude that the frequency of interruptions has increased over time, between the mid 1980s and early 1990s to the 2000's [4-9]. In these papers, the authors point to the increase in blackout frequency in the NERC System Disturbances Reports, without questioning whether this dataset is complete.

Findings or observations based on simple statistical analysis (e.g., mean, median) would potentially change if the dataset were more complete and accurate. For instance, [9] observes that the median customer loss for U.S. and Canadian interruptions between 1990 and 2004 were, respectively, 67,765 and one. This vast difference could in part be explained by the different populations in the countries, but most likely there is under-reporting of events as well. However, it is impossible to hypothesize whether the statistical measures would be higher or lower if a more accurate and complete dataset were available.

In the second category, there have been several papers examining whether the bulk power system displays characteristics of self-organized criticality (SOC). Systems that display SOC have non-linear dynamics and tend to be close to major disruption even when in an average system state. The SOC literature relies in part on time-series analysis (the implications of which are discussed above) but also on analysis of frequency and magnitude of interruptions.

Table 4. Literature that uses NERC or DOE emergency disturbance data

Paper	Type of Analysis	Dataset
Carreras 2000 [11]	SOC	NERC 1993-1998
Chen 2001 [13]	SOC	NERC 1984-1999
Talkudar 2003[8]	Time-series	NERC 1984-2000
Carreras 2004 [12]	SOC	NERC 1984-1998
Weron 2005 [14]	SOC	NERC 1984-2000
Amin 2007 [5]	Time-series	NERC 1991-2005
Simonoff 2007 [9]	Time-series Simple statistical	NERC 1990-2004
Amin 2008 [4]	Time-series	NERC 1991-2006 DOE 2001-2005
Lyons 2008 [7]	Time-series	NERC 2002-2007
Hines 2009 [10]	Time-series SOC	NERC 1984-2006
Amin 2011 [6]	Time-series	NERC 1995-2005 DOE 2001-2005

[11] and [12] both analyze a time series of NERC reliability data for evidence of SOC in bulk power interruptions. The authors of these papers examine long-term correlation of interruptions for evidence of SOC characteristics. These authors recognized that "[i]t is not clear how complete

this [NERC] data is, but it is the best-documented source that we have found for blackouts in the North American power transmission system." [12]

[13] presents a similar analysis as [11], but with an extended time-frame of data. [14] also examines SOC of interruptions, but arrived at a different conclusion—that SOC is not an appropriate model for power systems—using the same NERC data.

The SOC studies include some time-series analysis, specifically investigation of time between interruptions [10-14]. For SOC systems, probability of an interruption increases exponentially with the waiting time since the last interruption. Some authors found exponential decay in the inverse probability of waiting time [11-13], and some did not [14]. This type of analysis depends on the completeness of the dataset, and incomplete interruption data could affect the accuracy of the findings. If the data are incomplete, time between interruptions could appear longer than in a complete dataset, which suggests a quicker drop-off in the probability may better reflect reality. This may strengthen the argument for SOC in bulk power systems.

The SOC literature includes other kinds of analyses, including examination of the probability of interrupted customers, and frequency of interruptions based on number of customers affected [11-14]. Analysis of interruption frequency may be skewed by an incomplete dataset: a more complete dataset would contain more events, and therefore a higher frequency of events. However, it is not clear what the size of these missing (or incorrectly reported) events are, and, hence, how their inclusion would change the finding of SOC characteristics.

Our analysis of existing literature is a preliminary assessment. We cannot say with certainty whether and how the findings would change with a different dataset. This is merely a sketch of why having accurate and complete data is important for analysis of bulk power system reliability.

5. Suggestions for Improving the Completeness and Accuracy of Bulk Power System Reliability Data

Because statistical analysis of bulk power system reliability based on complete and accurate data is important, changes in the current data collection could provide much value to the electricity sector. In addition to the existing emergency reporting, which is vital for immediate response to events, we recommend an additional mandatory reporting mechanism that would collect final information about reliability events. This would provide a solid basis for analysis, for improved policy-making, and, as a result, for a more reliable bulk power system. In addition, having multiple sources of reliability data containing potentially redundant information is useful because they would offer researchers a way to validate, or at least calibrate, the results of reliability analysis from one source against another.

One way to collect more complete and accurate data would be to expand the existing emergency reliability reporting structure, but require an additional "archival" report with a much longer reporting time (on the order of weeks not hours). As evident in the examination of NERC detailed reports in section 3.3, when more time is allowed and a careful investigation is undertaken, it is possible to correct and update the information on number of customers interrupted, start time and duration, and possibly add new information not available at the time of the initial report. This reporting could be similar to or perhaps a part of NERC new Event Analysis initiative.[9]

Another possible way to develop a statistically-valid reliability dataset would be to require SAIDI and SAIFI reporting from all utilities or other service providers that serve end-use customers. Reliability metrics that measure customer interruptions can be satisfactorily calculated with data from these entities alone, since the focus here is on reliability events that result ultimately in interruptions of power to end-use customers. Requiring a report from service providers of annual SAIDI and SAIFI, distinguishing between the origin of the events they report on (e.g., bulk power versus distribution system). This would have the added benefit of avoiding the double-reporting that can arise when multiple layers of entities report, as happens often with existing NERC and DOE emergency reliability reporting. This would, however, not be adequate for the NERC and DOE emergency reporting that track many kinds of reliability events, not just ones that result in customer interruptions.

The frequency, detail and granularity of whatever new reliability reporting is developed needs to support the goal of the desired analysis. As noted above, one important measure of bulk power system reliability is number, frequency and duration of average customer interruptions. For that measure of reliability, an annual SAIDI and SAIFI value reported by utilities would be sufficient. However, for the SOC analysis reviewed above, researchers need to know the average length of time between outages - information not conveyed in annual SAIDI and SAIFI.

6. Conclusion

Reliability of the bulk power system is of vital importance to the economic and general welfare of the U.S. Knowing how reliable the system is, and how that reliability changes over time, is critical for developing and assessing technologies and policies that affect reliability. Currently, however, we lack a source of reliability data that is complete and accurate, and appropriate for use in analysis of trends or impacts of policies and technologies.

In this chapter we begin to assess the bulk power system emergency reliability reporting to NERC and to DOE. We find evidence that these two datasets may not be accurate and complete. This places in question results of reliability analysis conducted with these data. We critically review the existing literature that analyzes these data, and hypothesize how the inaccuracy and incomplete nature of the data may affect the findings. Finally, we present some thoughts on what kind of data collection activity may improve the available reliability data.

7. Acknowledgments

The research described in this chapter was sponsored by the U.S. Department of Energy Office of Electricity Delivery and Energy Reliability.

References

[1] DOE, "OE-417 Electric Emergency Incident and Disturbance Report", revised January 2011.
[2] NERC, "Standard EOP-004-1 --- Disturbance Reporting," January 2007.
[3] IEEE Power Engineering Society. 2004. IEEE Std 1366"-2003 IEEE Guide for Electric Power Distribution Reliability Indices. ISBN 0-7381-3890-8 SS95193. New York: Institute of Electrical and Electronics Engineers, Inc. May 14. 35 pages.
[4] S.M. Amin, "For the Good of the Grid", IEEE Spectrum Magazine, 6(6), November/December 2008, pp 48-59.
[5] S.M. Amin and P. F. Schewe, "Preventing Blackouts", Scientific American, May 2007, pp. 60-67.

[6] S.M. Amin, "U.S. Electrical Grid Gets Less Reliable", IEEE Spectrum, January 2011.
[7] C. Lyons, J. Jacobi, and R. Starkweather. "NERC Standards and Standards Compliance: Still a Work in Progress?", The Electricity Journal, 21(3), April 2008, pp. 29-39.
[8] S.N. Talkudar, J. Apt, M. Ilic, L.B. Lave, M.G. Morgan, "Cascading Failures: Survival versus Prevention", The Electricity Journal, 16(9), November 2003, pp. 25-31.
[9] J.S. Simonoff, C.E. Restrepo, R. Zimmerman, "Risk-Management and Risk-Analysis-Based Decision Tools for Attacks on Electric Power", Risk Analysis, 27(3), June 2007, pp. 547-570.
[10] P. Hines, J. Apt, and S. Talkudar, "Large Blackouts in North America: Historic trends and Policy implications", Energy Policy, 37(12), December 2009, pp. 5249-5259.
[11] B.A. Carreras, D.E. Newman, I. Dobson, and A.B. Poole, "Evidence for Self-Organized Criticality in a Time Series of Electric Power System Blackouts", Proceedings of the 33th Hawaii International Conference on System Science, January 2000.
[12] B.A. Carreras, D.E. Newman, I. Dobson, and A.B. Poole, "Evidence for Self-Organized Criticality in a Time Series of Electric Power System Blackouts", IEEE transactions on Circuits and Systems I: Regular Papers, 51(9), September 2004, pp 1733-1740.
[13] J. Chen, J.S. Thorpe, and M. Parashar, "Analysis of Electric Power System Disturbance Data", Proceedings of the 34th Hawaii International Conference on System Science, January 2001.
[14] R. Weron and I. Simonsen, "Blackouts, risk, and fat-tailed distributions", Proceedings of the 3rd Nikkei Econophysics Symposium, Springer-Tokyo, 2005/2006.

End Notes

[1] http://www.oe.netl.doe.gov/oe417.aspx (Accessed August 2011).
[2] This will likely change to 72 hours starting in 2013.
[3] Discussion with former EIA staff.
[4] Discussion with OE staff.
[5] http://www.nerc.com/page.php?cid=5|66 (Accessed August 2011).
[6] For events that fall under the NERC EOP-004 standard but do not trigger OE-417 reporting, the NERC Disturbance Report Form must be submitted to NERC.

[7] Some Canadian and Mexican utilities report to NERC but not DOE; Puerto Rico utilities report to DOE but not NERC.

[8] Two events with both emergency and detailed NERC reports were not included here because the emergency reports had very little information (e.g., no amount of generation lost, no customers interrupted).

[9] In October 2010 NERC began a pilot of this initiative, which requires either a Brief or Event Analysis report on all events within10 to 120 days. The template for the Phase 1 and Phase 2 field trials includes customer interruptions, as well as other information.

In: Electricity Reliability in the United States
Editor: Emmanuel Martinez
ISBN: 978-1-62417-149-9

Chapter 3

TRACKING THE RELIABILITY OF THE U.S. ELECTRIC POWER SYSTEM: AN ASSESSMENT OF PUBLICLY AVAILABLE INFORMATION REPORTED TO STATE PUBLIC UTILITY COMMISSIONS*

Joseph H. Eto and Kristina Hamachi LaCommare

ABSTRACT

Large blackouts, such as the August 14-15, 2003 blackout in the northeastern United States and Canada, focus attention on the importance of reliable electric service. As public and private efforts are undertaken to improve reliability and prevent power interruptions, it is appropriate to assess their effectiveness. Measures of reliability, such as the frequency and duration of power interruptions, have been reported by electric utilities to state public utility commissions for many years. This study examines current state and utility practices for collecting and reporting electricity reliability information and discusses challenges that arise in assessing reliability because of differences among these practices. The study is based primarily on reliability information for 2006 reported by 123 utilities to 37 state public utility commissions.

---

* This is an edited, reformatted and augmented version of Ernest Orlando Lawrence Berkeley National Laboratory publication LBNL-1092E, dated October 2008.

Acronyms and Abbreviations

ASAI	Average Service Availability Index
CAIDI	Customer Average Interruption Duration Index
CD	Census Division
DAWG	Disturbance Analysis Working Group
DC	District of Columbia
DOE	U.S. Department of Energy
EIA	Energy Information Administration
IEEE	Institute of Electrical and Electronics Engineers
IOU	investor-owned utility
LBNL	Lawrence Berkeley National Laboratory
MAIFI	Momentary Average Interruption Frequency Index
MED	major event day
MW	megawatt (10^6 watts)
NARUC	National Association of Regulatory Utility Commissioners
NERC	North American Electric Reliability Corporation
NRRI	National Regulatory Research Institute
OE	DOE Office of Electricity Delivery and Energy Reliability
PUC	Public Utility Commission
SAIDI	System Average Interruption Duration Index
SAIFI	System Average Interruption Frequency Index

Executive Summary

Large blackouts, such as the August 14-15, 2003 blackout in the northeastern United States and Canada, focus attention on the importance of reliable electric service.

As public and private efforts are undertaken to improve reliability and prevent power interruptions, it is appropriate to assess their effectiveness. Measures of reliability, such as the frequency and duration of power interruptions, have been reported by electric utilities to state public utility commissions (PUCs) for many years.

This study examines current state and utility practices for collecting and reporting electricity reliability information and discusses challenges that arise in assessing reliability because of differences among these practices.

To collect information on current practices and rules that guide utility-reported reliability information, we contacted all 50 state PUCs as well as the District of Columbia (DC) PUC.

When permitted by state practices, we also collected a large sample of publicly available, actual reliability information reported by utilities to the PUC for year 2006. In total, we received information provided by 123 utilities to 37 state PUCs (see Figure ES-1). In aggregate, the reliability information we collected represents over 77% of total electricity sales by state-regulated investor-owned utilities or nearly 60% of total U.S. electricity sales.

Our assessment focused on three reliability metrics: System Average Interruption Duration Index (SAIDI), System Average Interruption Frequency Index (SAIFI), and Momentary Average Interruption Frequency Index (MAIFI). SAIDI and SAIFI measure the duration and frequency, respectively, of sustained interruptions; MAIFI measures the frequency of momentary interruptions.

Taken together, these three metrics can be used to develop a comprehensive assessment of reliability nationwide.

Our findings regarding state PUC practices and rules on reliability information reported by utilities are summarized as follows:

- Thirty-five state PUCs, including DC, require routine reporting of reliability event information. This is a net increase of 10 state PUCs over the number reported in a similar survey conducted by the National Regulatory Research Institute (NRRI) in 2004.
- These 35 PUCs require annual reporting of SAIDI and SAIFI and/or the Customer Average Interruption Duration Index (CAIDI), which, along with SAIFI, can be used to derive SAIDI. Only two state PUCs require reporting of MAIFI.
- Twenty-one PUCs have reporting requirements that formally define major events. Of these 21, four require reporting following the Institute of Electrical and Electronics Engineers (IEEE) Standard 1366-2003, IEEE Guide for Electric Power Distribution Reliability Indices, which introduces a consistent means for defining major events using the concept of “major event days.”
- An additional four PUCs receive reliability information from utilities, though not as a result of a formal reporting requirement.
- Thirty-seven state PUCs, including DC, make publicly available or summarize in publicly available documents, the reliability information they collect from utilities.

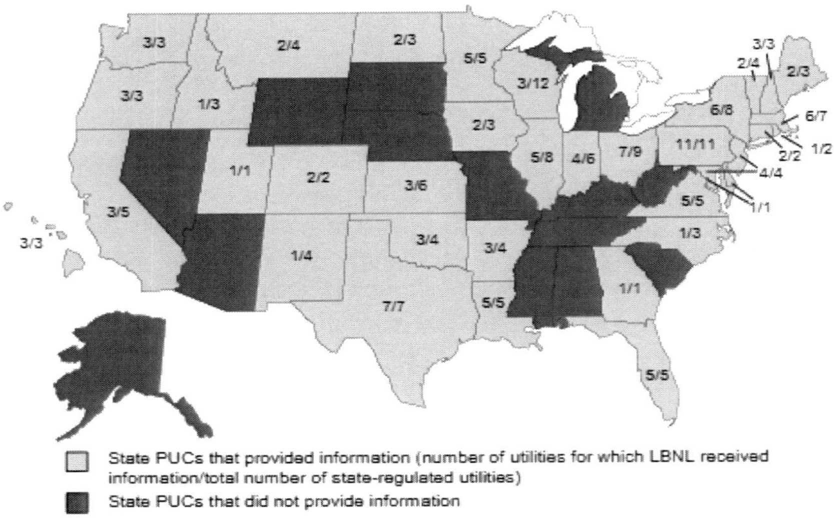

Figure ES- 1. Summary of States that Provided Utility-Reported Reliability Information.

Our findings regarding utility practices for collecting and reporting reliability information to state PUCs are summarized as follows:

- All utilities reported SAIDI and SAIFI (and/or CAIDI). Only 12 of the 123 utilities reported MAIFI.
- Summary statistics for reported SAIDI, SAIFI, and MAIFI exhibit observable though not statistically significant variations across census regions.
- The definition of and practices for recording sustained and momentary interruptions have evolved over time leading to inconsistencies among utilities.
- Differences in the definition of a sustained interruption do not appear to affect SAIDI or SAIFI in a statistically significant manner.
- Utilities define major events as a means for distinguishing between utility performance in planning for and responding to routine interruptions versus that for non-routine or extraordinary interruptions.
- The definition of a major event is not consistent among the majority of utilities.

- IEEE Standard 1366-2003 introduces a consistent means for defining major events using the concept of "major event days."
- Some utilities report SAIDI and SAIFI both including and not including major events; other utilities only report SAIDI and SAIFI not including major events.
- When major events are not included, SAIDI is lowered relatively more than SAIFI compared to when major events are included.
- Many utilities report descriptive information on each major event.
- Use of IEEE Standard 1366-2003 does not appear to bias SAIDI or SAIFI values compared to using prior definitions of major events.

We also collected information on bulk power system emergencies reported by utilities in near real-time to national bodies in 2006, including the U.S. Department of Energy (DOE) and the North American Electric Reliability Corporation (NERC), and compared aspects of this information to that reported by utilities to state PUCs. Our findings are summarized as follows:

- Information on electricity reliability reported to these two national bodies consists of descriptive information that is reported in near real-time on individual, large events that affect the bulk power system. The reporting takes place in near real-time because an important purpose of the reporting is to notify relevant industry and public bodies of significant power system events that may require immediate response. With few exceptions, the same information is reported to both DOE and NERC at the same time.
- Many, but not all, events reported to these national bodies also cause power interruptions to customers. For these events, the number of customers affected is reported.
- An initial assessment of these events supports the conventional wisdom that the majority of power interruptions experienced by customers are not due to large events that affect the bulk power system; they are due to more localized events that affect only utility distribution systems.
- It is difficult to cross-reference information reported to national bodies on individual large bulk power system events that cause power interruptions, as defined by these national bodies, with information reported to state PUCs on individual major events, as defined by either the PUC or the reporting utility.

From these findings, we draw the following conclusions and recommendations:

- State PUC interest in electricity reliability is growing.
- However, differences in utility reporting practices hamper meaningful comparisons of reliability information reported by utilities to different state PUCs and, therefore, may limit the effectiveness of efforts to measure the effectiveness of efforts to improve reliability.
- Efforts to eliminate differences that are solely due to reporting practices are just beginning. These efforts, which focus on using standard definitions, such as those promoted by IEEE Standard 1366-2003, are promising and should be encouraged.
- Until IEEE Standard 1366-2003 is adopted universally, regulators concerned about the definition and treatment of major events in reporting reliability information should consider requiring reporting of SAIDI and SAIFI both including and not including major events, as well as descriptive information on each major event.
- More work is required to better understand the sources of discrepancies and the importance of seeking greater consistency between reliability information reported to national bodies and that reported to state PUCs.

1. INTRODUCTION

Large blackouts, such as the August 14-15, 2003 blackout in the northeastern United States and Canada, focus attention on the importance of reliable electric service. Acknowledged as the largest power outage ever to occur in North America, the 2003 blackout affected more than 50 million people in eight U.S. states and two Canadian provinces; some customers were without power for more than two days. The final blackout investigation report contained 46 recommendations to improve electricity system reliability and prevent future blackouts (Department of Energy 2004).

As a result of this major blackout, public and private efforts have been proposed to improve reliability and prevent future power interruptions, both large and small. It is important to assess the effectiveness of these efforts. A common management precept is that you cannot manage something effectively unless you can measure it. Performance metrics, similar to the letter grades assigned in school, are a way to quantitatively measure reliability and improvements in it.

Lagging measures of reliability, so called because they are retrospective, have been recorded by electric utilities for many years. These metrics, such as the frequency and duration of power interruptions, have been an essential tool for managing reliability because they provide a quantitative, objective basis for judging the effectiveness of the organization's efforts to maintain or improve reliability. These data have supported organizations' and regulators' efforts to monitor reliability performance and to compare performance to benchmarks or trends, so that these entities can initiate corrective actions when necessary to improve performance.

This study presents findings on current state and utility practices for collecting and reporting reliability information and discusses challenges that arise in assessing reliability as a result of differences among these practices. We pay special attention to the effect of the recent adoption of the Institute of Electrical and Electronics Engineers (IEEE) Standard 1366-2003, IEEE Guide for Electric Power Distribution Reliability Indices (IEEE Power Engineering Society 2004), which formally defines the major reliability metrics and the factors and procedures used to calculate them (see Text Box).

Although a variety of reliability metrics exist, we focus primarily on practices associated with two of the most commonly reported metrics, the System Average Interruption Duration Index (SAIDI) and the System Average Interruption Frequency Index (SAIFI). Both are used to describe interruptions of a defined minimum duration. We also focus on a less commonly reported metric, the Momentary Average Interruption Frequency Index (MAIFI),[1] which captures interruptions that are shorter in duration than those recorded for SAIDI and SAIFI. Taken together, the three metrics can be used to comprehensively assess reliability nationwide (LaCommare and Eto 2004).

To conduct the current study, we contacted public utility commissions (PUCs) in all 50 states plus the District of Columbia (DC) to gather information on practices and rules related to utility-reported reliability information for year 2006. When permitted by state practices, we collected a large sample of publicly available reliability information reported by utilities for 2006. We chose to focus on state PUCs and state-regulated investor-owned utilities because many state PUCs already routinely collect this information from the utilities they regulate and because these utilities, taken together, account for a significant share of total sales of electricity in the U.S.

We first developed descriptive statistics, organized by Census Division regions, on the information reported by utilities. We then examined the impact of differences in utility reporting practices on the reliability indices.

We also collected information on electricity reliability reported by utilities to national bodies on large power system emergencies that sometimes cause power interruptions, including both the U.S. Department of Energy (DOE), and the North American Electric Reliability Corporation (NERC), and compared aspects of this information to the information reported by utilities to state PUCs.

The remainder of this chapter is organized as follows:

- Section 2 describes the rationale for and the results of our contacts with state PUCs.
- Section 3 presents findings on state PUC practices and rules that affect reliability information reported by utilities.
- Section 4 presents qualitative and quantitative findings on utility practices for collecting and reporting reliability information to state PUCs.
- Section 5 presents findings from efforts to compare information on bulk power system electricity emergencies that cause power interruptions and are reported in near real-time to national bodies to the reliability information provided to state PUCs on a routine basis.
- Section 6 summarizes all of our findings and conclusions.

IEEE Standard 1366-2003

The IEEE established the Standard 1366 for defining reliability indices in order to:

"present a set of terms and definitions which can be used to foster uniformity in the development of distribution service reliability indices, to identify factors which affect the indices, and to aid in consistent reporting practices among utilities" (IEEE Power Engineering Society 2004).

IEEE Standard 1366-2003 is a voluntary means to derive consistent reliability metrics that can be effectively evaluated for decision making and policy-making purposes (Gonzalez 2006). It provides definitions for SAIDI, SAIFI, CAIDI, etc including methods for calculating these statistics.

The goal is for all utilities and entities that need a methodology for estimating reliability at the distribution level to adopt this standard so that all utilities have a consistent benchmark for comparison and reporting across the U.S. (Warren 2006).

IEEE Standard 1366-2003 is an update of previous versions (i.e., 1366-1998 or 1366-2001), which is designed to define in a consistent manner reliability indices as well as major events that affect utility distribution service territories.

The IEEE Standard 1366-2003 standard defines a new method for identifying major events, called the "2.5 beta method." A Major Event Day is defined as a day in which the daily system SAIDI exceeds a threshold value, T_{MED}. The threshold value is calculated using statistical criteria to identify events that are significantly different from the majority of events experienced in recent years (up to five sequential years of historical data). The methodology is described in greater detail in Appendix A.

2. The Rationale for, and Process and Results of Contacts with State PUCs

To conduct this study, we contacted all 50 state PUCs plus DC to gather information on current practices and rules that guide utility-reported reliability information. When permitted by state practices, we also collected actual reliability information reported by utilities for year 2006. This section describes the rationale for, questions asked during, and results of our contacts with state PUCs.

2.1. Why Focus on State PUCs and State-Regulated Utilities?

We chose to contact state PUCs for several reasons. First, the utilities they regulate, investor-owned utilities (IOUs), collectively account for the majority of total U.S. electricity sales (nearly 75%). Second, state PUC oversight of these utilities relies on information that is routinely provided by the regulated utility to the PUC, including information on reliability. Third, state PUCs make much of the information they receive from the regulated utility publicly available, pursuant to administrative rules established by the state.

We anticipate future work that will collect and assess publicly available utility-reported reliability information from other sources, such as utility industry trade associations, federal utility lending agencies, and of course directly from individual utilities.

2.2. What Did We Ask Each State PUC?

We asked each PUC the following questions:

- Does your state require utilities to submit reliability information to the PUC?
- If so, do the requirements call for routine (e.g., annual) submission of this information?
- What reliability information (e.g., indices) is submitted?
- How do reporting practices or rules guide the definition of sustained interruptions and major events used to calculate the major reliability indices? Specifically, has the state adopted the recent IEEE Standard 1366-2003, which formally defines these indices and how these factors should be used to calculate them?
- What are the practices or rules for reporting reliability indices with major events included and not included?
- What are the practices or rules guiding reporting on individual major events?
- Does the PUC make publicly available either the reliability information provided to the PUC or summaries of this information?
- If so, can you provide our research team with whatever utility-reported reliability information is readily available?

The information we collected reflects state PUC practices that were current in 2006.

2.3. What Utility-Reported Reliability Information Did We Collect?

We contacted every state PUC as well as DC to obtain information on current state reporting requirements.[2] The findings from this portion of our research are reported in Section 3. We were also able to obtain 2006 reliability information for 123 utilities from 37 PUCs including DC. These findings are reported in Section 4. Figure 1 shows the 37 states (including DC) that provided utility-reported reliability information. The fourteen states we were not able to obtain utility-reported reliability information are shown in dark gray shading. Figure 1 also tabulates the number of utilities for which we obtained reliability information (numerator) compared to the total number of regulated-investor-owned utilities within the state (denominator). In an effort to collect as much utility-reported reliability information as possible and at the same time respect the workload of state PUC staff, we requested only readily available information. In some cases, we received annual summaries that PUC staff had prepared by tabulating information from individual reports submitted by each utility in the state. In other cases, we received copies of some, but not necessarily all, of the individual reports submitted by the utilities.

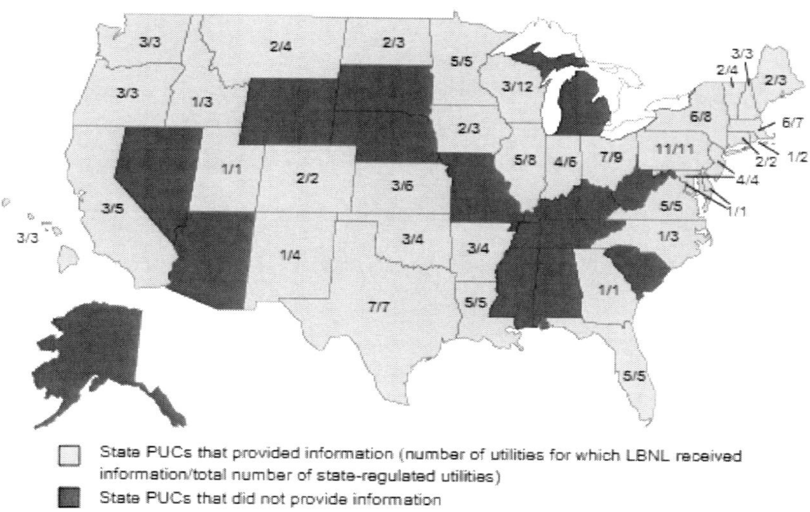

Figure 1. Summary of States that Provided Utility-Reported Reliability Information.

3. Review of State Reporting Requirements and Practices for Utility Reliability Information

We contacted PUCs in every state and DC to obtain information on current state reporting requirements for utility reliability information.[3] This section presents our findings on overall state reporting requirements and practices, changes in reporting requirements over time, and the specificity of state requirements regarding the types of reliability information that must be reported and the definitions of major events. The information we report reflects state practices, as of 2006.

3.1. Requirements for Routine Reporting and Practices for Public Availability of Reported Information

We find that 35 PUCs, including DC, require that reliability information be reported routinely. An additional four state PUCs receive reliability information from utilities though not in response to a formal reporting requirement.

Thirty-seven state PUCs, including DC, make publicly available or summarize in publicly available documents the reliability information they collect from utilities.

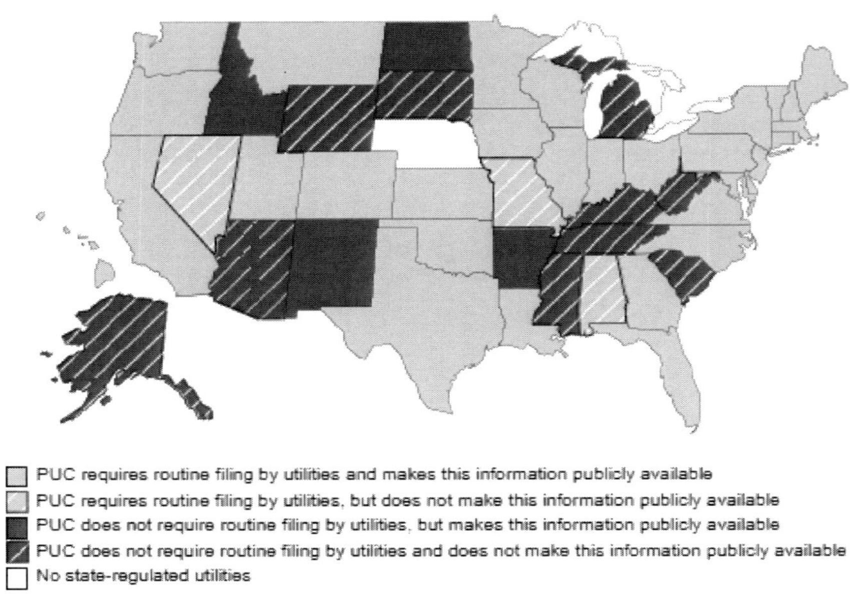

Figure 2. State Reporting Requirements and Practices for Utility-Reported Reliability Information.

Figure 2 combines information on state reporting requirements with information on state practices regarding the public availability of reported information. The 33 states shown in light solid shading, including DC, currently require utilities to report reliability information and make this information publicly available (either the actual reports submitted by the utilities or summaries prepared by PUC staff). The four states denoted by solid dark shading do not require utilities to report reliability information, yet did have information they were willing to share with us.[4]

The 13 states shown in dark and light-striped shading either do not require or do not receive utility-reported reliability information and do not make it available publicly if they receive it. The 10 states shown in dark-striped shading (Alaska, Arizona, Kentucky, Michigan, Mississippi, South Carolina, South Dakota, Tennessee, West Virginia, and Wyoming) do not currently require utilities to report reliability information. The three states denoted by light-striped shading (Alabama, Arkansas, and Nevada) require utilities to report reliability information but do not make this information publicly available.

As noted previously, Nebraska, shown in white, has no state-regulated utilities.

3.2. Changes in Requirements for Reporting Over Time

In 2004, the National Regulatory Research Institute (NRRI), the research branch of the National Association of Regulatory Utility Commissions (NARUC), surveyed state PUC requirements and practices for utility-reported reliability information. By comparing the findings from the NRRI survey to those from the current work, we can begin to assess changes in state PUC reporting requirements over time.

As summarized in Table 1, NRRI received responses from 42 PUCs. Of these, 25 PUCs had routine reporting requirements for utility reliability information. Our current research finds that, in 2006, 35 PUCs required routine reporting, a net increase of 10 PUCs compared to the NRRI survey.

It is important, however, to qualify this apparent increase as follows. A total of 23 PUCs that required routine reporting in 2006 were also found by NRRI to require routine reporting in 2004, while two PUCs that NRRI found required routine reporting no longer require this reporting (Arkansas and Idaho). Seven PUCs that required routine reporting in 2006 were not found by

NRRI to require routine reporting in 2004. Five PUCs that required routine reporting in 2006 were PUCs that did not respond to the NRRI survey in 2004.

Table 1. Assessment of State PUC Reporting Requirements Over Time

	PUCs that Indicated Routine Reporting Requirements in 2004 NRRI Survey	PUCs that Confirm Routine Reporting Requirements in LBNL 2006 Study
Yes	25	35
No	17	16
Did Not Respond	9	0

3.3. Reporting Requirements for Metrics and the Definition of Major Events

In Section 4, we discuss the reliability information reported by utilities to state PUCs. That discussion points to important differences in reporting practices that complicate review and assessment of information reported by utilities in different states. In this sub-section, we lay the groundwork for that discussion by reviewing part of the basis for these differences, which is found in the specificity of state requirements regarding the type of reliability information that must be reported and the definition of major events.

3.3.1. Reliability Metrics Reported

The reported reliability metrics do vary among the states that require routine reporting. However, the two most common measures of reliability (SAIDI and SAIFI) are generally consistent across all states (see Figure 3).

All 35 states require reporting of SAIFI. Thirty states require reporting of SAIDI and 18 states require reporting of the Customer Average Interruption Duration Index (CAIDI). However, the definitions of SAIDI, SAIFI, and CAIDI are interrelated. SAIDI can be calculated from SAIFI and CAIDI and CAIDI can be calculated from SAIDI and SAIFI (see Appendix A). Therefore, for all intents and purposes, SAIDI and SAIFI are either directly reported or, if SAIDI is not directly reported, it can be derived from SAIFI and CAIDI, which are reported.

In contrast, only two states require reporting of MAIFI and only three states require reporting of the Average System Availability Index (ASAI), an index that can be derived from CAIDI.

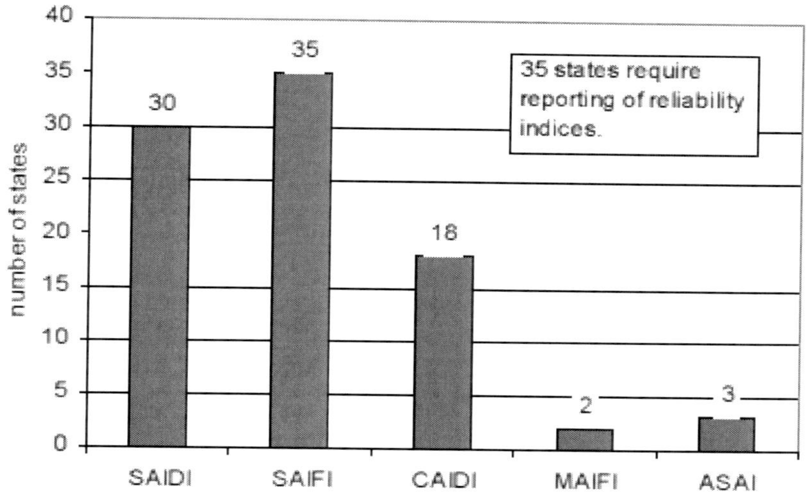

Figure 3. State Reporting Requirements for Reliability Metrics.

3.3.2. Definition of Major Events

Information on reliability is sometimes segmented using the concept of major events.

Major events are defined by a variety of criteria to differentiate between routine power interruptions and non-routine or extraordinary power interruptions.

In Section 4, we review specific variations in the way major events are defined and assess the impact that segmenting reliability information using these definitions has on reliability metrics.

Here, we report the extent to which state requirements determine utility practices for defining major events.

Of the 37 PUCs that made utility-reported reliability information available for our study, 21 said they have adopted a formal definition of a major event. Of these 21, four states (Colorado, Delaware, DC, and Utah) require the use of the IEEE Standard 1366-2003 to define major events using the concept of major event days (see Text Box on page 2).

The remaining 17 states use some other definition for a major event. Fifteen states do not formally define major events. We were not able to obtain information from one state about whether it has adopted a formal definition of major events.

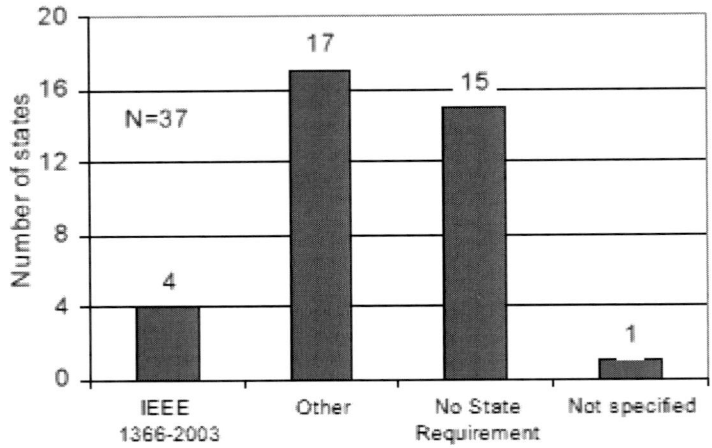

Figure 4. State Reporting Requirements for the Definition of Major Events.

4. Review and Assessment of Utility-Reported Reliability Information to State PUCs

We received utility-reported reliability information for 123 utilities from 37 state PUCs representing year 2006. In this section, we present qualitative and quantitative findings and assess utility practices for collecting and reporting reliability information to state PUCs. Our assessment focuses on three reliability metrics, SAIDI, SAIFI, and MAIFI, which, taken together, can be used to comprehensively assess reliability on a national basis (LaCommare and Eto 2004).

We begin by reviewing reliability information aggregated by U.S. Census Division regions (Energy Information Administration 2003). See Figure 5. Although we observe discernable trends in the aggregated information, the process of aggregating made us aware of differences among utility practices for defining and presenting reliability information. Without a better understanding of the impact of these differences on the reported information, it is difficult to ascribe the trends we observe to material differences in reliability among utilities.

Thus, the majority of this section reviews and attempts to quantify the impact of differences in reporting practices on reported reliability metrics. We focus initially on differences in practices for defining sustained interruptions.

We then focus on practices regarding the treatment of major events, including the practice of reporting reliability metrics including and not including major events, the definitions of major events, and the effect of adopting IEEE Standard 1366-2003, which provides a standardized, statistically based method of defining major events.

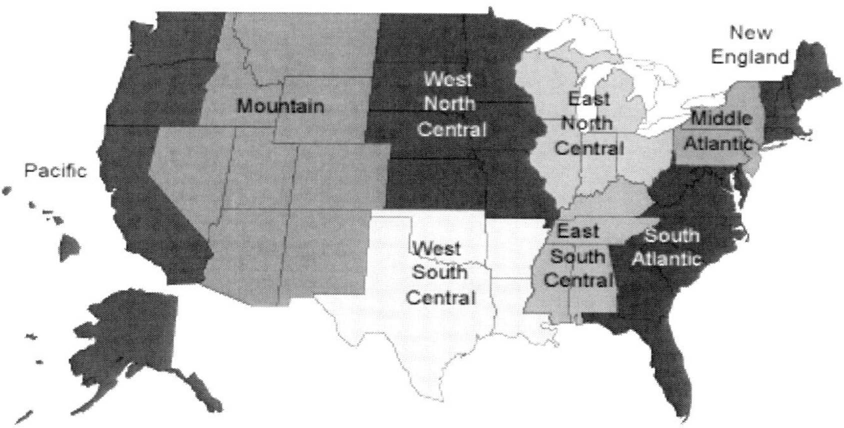

Figure 5. Map of U.S. Census Divisions.

4.1. Regional Trends in SAIDI, SAIFI, and MAIFI

The data in Table 2 represent summary statistics (number of observations, simple average, and standard deviation) for the reliability information (SAIDI, SAIFI, and MAIFI) we received aggregated to regions as defined by the nine U.S. Census Divisions. Table 2 also provides two indicators of the representativeness of the information we received for each region: 1) the electricity sales by the utilities for which we have reported information as a percentage of the total electricity sales by all state-regulated investor-owned utilities within that region; and 2) the electricity sales by the utilities for which we have reported information as a percentage of the total electricity sales by all electric utilities (both state-regulated and non-state-regulated) within that region. Information on electricity sales is taken from EIA Form 861 (Energy Information Administration 2006).

As reflected in Table 2, the information we collected accounts for over three-quarters of total electricity sales by state-regulated utilities or nearly 60% of total U.S. electricity sales. However, the representativeness of the information we collected varies by region. For some regions (Middle Atlantic, New England, and Pacific), we received information representing essentially all electricity sales by state-regulated utilities within these regions. For other regions (East North Central, South Atlantic, and West South Central), we received information representing more than 70% of electricity sales by state-regulated investor-owned utilities within these regions. For the remaining regions (Mountain, West North Central, and East South Central) we received information representing progressively lower proportions of electricity. For one region, East South Central, we did not receive information for any of the ten state-regulated utilities within this region.

All 123 utilities reported SAIDI and SAIFI (and/or CAIDI, which along with SAIFI can be used to calculate SAIDI). Only 12 of the 123 utilities reported MAIFI.

In general, one might expect reliability to vary because of regional differences in climate, vegetation, and population. Visual review of the information we collected averaged by region shows greater variation in SAIDI than in SAIFI. A greater than two-to-one difference can be seen in average SAIDI, ranging from a low of 118 minutes in the Mountain region to a high of 498 minutes in East North Central. Variations in average SAIFI are smaller, ranging from a low of 1.22 in Mountain to a high of 1.99 in Pacific. However, despite the ranges observed in average SAIDI and SAIFI, the high standard deviations associated with them indicate that the differences among regions are not statistically significant.

Similarly, the very small number of reported MAIFI values makes it difficult to ascribe a statistical significance to even the wide range of observed variation in these values. Accordingly, the remainder of this review will largely focus on SAIDI and SAIFI.

The significance of differences among the summary statistics emerging from this initial review is further diminished by the recognition that utility practices in collecting and reporting these reliability metrics vary.

4.2. Utility Practices for Defining Sustained Interruptions

Taken together, SAIDI, SAIFI, and MAIFI provide a comprehensive measure of electricity reliability because they capture the key features of the range of power interruptions experienced by electricity consumers. The SAIDI and SAIFI indices measure the duration and frequency, respectively, of sustained interruptions, while MAIFI measures the frequency of momentary interruptions. The definitions for sustained and momentary interruptions are interdependent, whereby a sustained interruption is any interruption that is not classified as a momentary event.

Although these definitions ensure that all power interruptions are classified as either sustained or momentary, utility definitions of whether an interruption is sustained or momentary vary. Figure 6 shows the different definitions for sustained interruptions that we found in the utility-reported reliability information we received.

Consistent with IEEE Standard 1366-2003 (and prior versions of this standard), the most common definition of a sustained interruption is one that is greater than or equal to 5 minutes in duration. Still, a fair number of utilities use other definitions. For example, nearly a quarter of the utilities (28) use a shorter duration (one or two minutes) for sustained interruptions.

Table 2. Summary of Utility-Reported SAIDI, SAIFI, and MAIFI by Census Division

Census Division	Sales as Percentage of Total IOU Sales in Region	Sales as Percentage of Total U.S. Sales in Region	SAIDI (minutes)			SAIFI			MAIFI		
			N	Avg	Std Dev	N	Avg	Std Dev	N	Avg	Std Dev
New England	99%	68%	16	198	130	16	1.44	0.62	ND	ND	ND
Middle Atlantic	100%	75%	21	225	188	21	1.28	0.55	ND	ND	ND
East North Central	75%	62%	19	498	895	19	1.46	0.48	ND	ND	ND
West North Central	57%	35%	12	166	202	12	1.31	0.68	2	5.11	5.03
South Atlantic	71%	53%	18	320	200	18	1.86	0.62	4	11.1	2.16
East South Central	0%	0%	ND	ND	ND	ND	ND	ND	ND	ND	ND
West South Central	88%	30%	18	134	56	18	1.38	0.46	ND	ND	ND
Mountain	35%	27%	7	118	58	7	1.22	0.54	ND	ND	ND
Pacific	99%	62%	12	296	214	12	1.99	1.21	6	3.40	2.35
U.S.	77%	58%	123	244	243	123	1.49	0.64	12	6.55	3.18

Note: N = number of reported values; Avg = average; Std Dev = standard deviation; ND = no data.

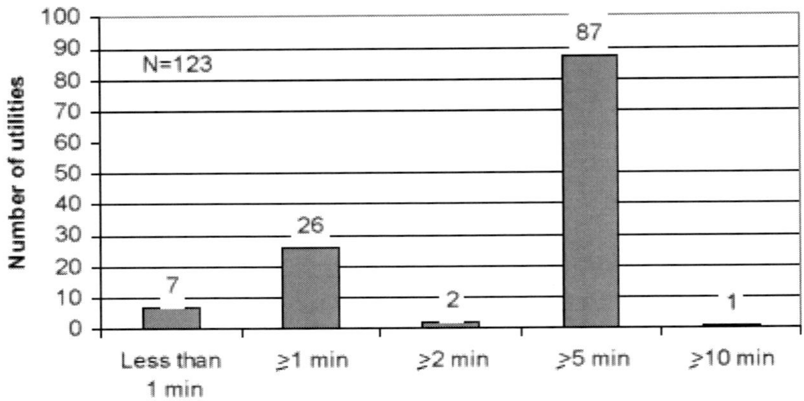

Figure 6. Utility Practices for Defining Sustained Interruptions.

First principles suggests that differences in utility practices regarding the minimum duration used to classify an interruption as sustained would tend to affect SAIFI more than they would affect SAIDI. We examined this suggestion by comparing the reliability metrics reported by the two largest subsets of utilities grouped according the definitions they used to specific the duration of a sustained interruption.

Table 3 compares SAIDI and SAIFI (both not including major events) for the 76 utilities that define a sustained interruption as lasting 5 minutes or longer to the 19 utilities that define a sustained interruptions as lasting 1 minute or longer. We find that these differences in definitions do not represent, by themselves, statistically significant differences in reported SAIDI or SAIFI between these two groups.

4.3. The Origins of MAIFI and the Significance of Momentary Interruptions

In discussing with industry experts utility definitions of sustained and momentary interruptions, we find that momentary interruptions are a comparatively recent concern for utilities and that practices for recording them are not widespread. The underlying factors include technical features now inherent in the design and operation of modern utility electricity delivery systems, increased sensitivity of certain end use devices to momentary interruptions, and the difficulty of reliably tracking momentary interruptions.[5]

Table 3. Summary of Utility-Reported SAIDI and SAIFI With Major Events Not Included for Utilities Using 1-Minute and 5-Minute Definitions for Sustained Interruptions

	SAIDI		SAIFI	
	5 min without major events included	1 min without major events included	5 min without major events included	1 min without major events included
Average	165	143	1.3	1.4
Std Dev	97	63	0.5	0.7
Median	145	151	1.3	1.3
N	78	19	78	19

The ability of an electricity delivery system to ensure reliability, as measured by the continuity of electricity service, depends on both the design and operation of the system, as well as the external conditions (e.g., weather) in which the system is expected to operate. A key technical element of modern electricity delivery systems is automatic devices that protect systems from damage during a disturbance and then, following the disturbance, restore electricity service. These devices sense and isolate portions of a system automatically to prevent a disturbance (e.g., a lightning strike) from damaging the system. In so doing, the devices interrupt service to the portions of the system that are "downstream" of the disturbance. Because the majority of disturbances are momentary, the devices are also designed to reconnect the isolated portions of the systems (i.e., reclose the opened circuit) automatically after a pre-determined delay. If the disturbance has passed, the reconnection will be successful and service will be restored. If the disturbance has not passed, the device will again automatically re-isolate that portion of the system. This sequence may be repeated more than once. At some point, usually under one minute, the devices, following a pre-determined rule, will cease attempting to reconnect the isolated portion of the system. At this point, the utility must take additional steps to restore service to the isolated portion of the system.

The number of reconnection attempts that a device is allowed to make and the time-delay between attempts are choices specified by the utility when it selects the device. The number of customers that may be affected by the operation of this device is also determined by the utility in designing its electricity delivery system and locating these devices within the system. These choices are guided by common industry practices as well as by individual utility design philosophies.

The key point here is that the momentary interruption of electricity service associated with operation of these devices is an intentional feature of the design and safe operation of the majority of modern electricity delivery systems. Prior to the introduction of these devices on distribution circuits in the 1940's, all interruptions of electric service were sustained interruptions because utility efforts to restore service required manual actions by utility personnel.

The introduction of these devices enabled automatic, rather than time-consuming manual, restoration of electric service. As a result, the frequency and duration of sustained interruptions decreased.

Public interest in momentary interruptions began in the 1970's when digital, rather than analog, electricity consuming devices were introduced to consumers. Early versions of these devices lacked a back-up source of power and were, therefore, especially vulnerable to momentary interruptions. Within the residential sector, digital clocks would start blinking and need to be reset following a momentary interruption. Within the commercial and industrial sector, machinery and processes controlled by programmable logic chips would stop operating following a momentary interruption (or a power quality event, such as a voltage sag event).

In other words, momentary interruptions, which resulted intentionally from utility efforts to reduce sustained interruptions, were now being recognized as causing new reliability problems for certain types of electricity consuming devices and processes. Consequently, there was movement toward formally defining and recording momentary interruptions.

Differences of opinion, however, exist regarding utility versus customer responsibilities for addressing the impacts of momentary interruptions and, therefore, of the importance of tracking momentary interruptions. Some experts believe that the most cost-effective way to reduce customer impacts from momentary interruptions (and power quality events) is to reduce the vulnerability of the individual electricity consuming devices or processes that are affected, rather than undertake measures on the utility-side of the meter to reduce the frequency of momentary interruptions.

Finally, it is important to bear in mind that, until recently, the methods for recording the duration of interruptions, both sustained and momentary, have been largely manual. For sustained interruptions, rounding to the nearest minute, and starting with a minimum duration of say 5 or 10 (or even 30) minutes was common practice. For momentary interruptions, it is not usually possible, unless utility personnel happen to be present at the time to record when an automatic restoration device activates. Only with the advent of automated outage management and recording equipment has more precise tracking of the duration of interruptions (both sustained and momentary) and numbers of customers affected been possible.

Together, these factors help explain why so few utilities report MAIFI and why there are differences among utilities in defining the transition between momentary and sustained interruptions. Momentary interruptions are a feature of the intended operation of modern electricity delivery systems and there are differences of opinion as to whether they should be tracked in the same manner as sustained interruptions. Moreover, it is difficult to record momentary interruptions accurately without expensive monitoring equipment.

4.4. Utility Practices for Segmenting Reliability Information Based on Major Events

As noted in Section 3, major events are defined by a variety of criteria that differentiate between routine power interruptions and non-routine or extraordinary power interruptions. Utility practices vary regarding the treatment of major events in reporting reliability metrics, including SAIDI and SAIFI. Some utilities report reliability metrics with major events included; others report these metrics with major events not included. And some utilities report the metrics both with major events included and not included.

Table 2 was developed by averaging a single SAIDI and SAIFI value from each utility for which we had reported information. We sought first to include values reported with major events included. However, when a utility reported only values with major events not included, we used that value. Table 4 gives a more complete picture of all the reliability information we received as it presents summary statistics on SAIDI and SAIFI both with major events included and not included.

Table 4 indicates that we received more SAIDI and SAIFI values without major events included than with major events included. The averages of SAIDI and SAIFI with major events are generally larger than those with major events not included and the differences can be significant. For example, the average SAIDI with major events is more than twice the average SAIDI with major events not included for the East North Central region.

Higher averages for figures that include major events would appear to be a predictable result; i.e., if some events are not included, then the resulting values of SAIDI and SAIFI should be lower. However, this is not always the case. The average SAIDI with major events not included is higher than the average SAIDI with major events included for the West South Central region.

Table 4. Summary of Utility-Reported SAIDI and SAIFI With Major Events Included and Not Included

	SAIDI								SAIFI							
Census Division	With Major Events Included				With Major Events Not Included				With Major Events Included				With Major Events Not Included			
	N	Avg	Std Dev	Wgt Avg	N	Avg	Std Dev	Wgt Avg	N	Avg	Std Dev	Wgt Avg	N	Avg	Std Dev	Wgt Avg
New England	7	260	196	312	17	148	88	147	7	1.40	0.66	1.42	17	1.26	0.61	1.29
Middle Atlantic	7	399	239	566	21	156	97	214	7	1.54	0.73	1.03	21	1.13	0.44	0.96
East North Central	19	498	895	424	15	150	60	133	19	1.46	0.48	1.46	15	1.24	0.19	1.22
West North Central	6	256	263	131	12	107	84	92	6	1.58	0.87	1.17	12	1.25	0.64	1.06
South Atlantic	15	350	207	254	18	212	111	143	15	1.94	0.65	1.76	18	1.50	0.40	1.39
East South Central	ND	ND	ND	ND	ND	ND	ND	ND	ND	ND	ND	ND	ND	ND	ND	ND
West South Central	10	114	47	115	18	126	58	130	10	1.24	0.50	1.16	18	1.33	0.46	1.32
Mountain	5	126	68	139	5	112	60	117	5	1.22	0.61	1.35	5	1.20	0.58	1.47
Pacific	9	332	238	245	12	156	71	128	9	1.93	1.21	1.34	12	1.59	0.95	1.08
U.S.	78	292	269	234	118	146	79	144	78	1.54	0.71	1.02	118	1.31	0.53	1.13

Note: N = number of reported values; Avg = average; Std Dev = standard deviation; Wgt Avg = weighted average; ND = no data.

Although the number of utility reports contributing to these averages is identical, comparison between Table 4 and Table 2 confirms that different utilities comprise each of these averages.

Although there is some overlap between the utilities whose reported information contributes to both averages (i.e., utilities that report SAIDI and SAIFI both with major events included and not included), many reported values are also included from utilities that reported SAIDI and SAIFI either with major events included or with major events not included (but not both).

The above review makes clear that treatment of major events in SAIDI and SAIFI must be taken into account when reviewing utility-reported reliability information. We now focus on direct examination of the significance of differences in practices for reporting SAIDI and SAIFI.

4.5. Reporting SAIDI and SAIFI with Major Events Included and Not Included

As discussed above, it is difficult to assess the effect of reporting SAIDI and SAIFI with major events included and not included when the utility reports that represent this information do not all include both pieces of information. Fortunately, 71 of the 123 SAIDI and SAIFI values we received were reported both with and without major events included. By focusing on these 71 reports, we can directly examine the effect of this difference in reporting practices.

Figures 7 and 8 present the distributions of the reported values and summary statistics for these 71 utility reports of SAIDI and SAIFI, respectively.

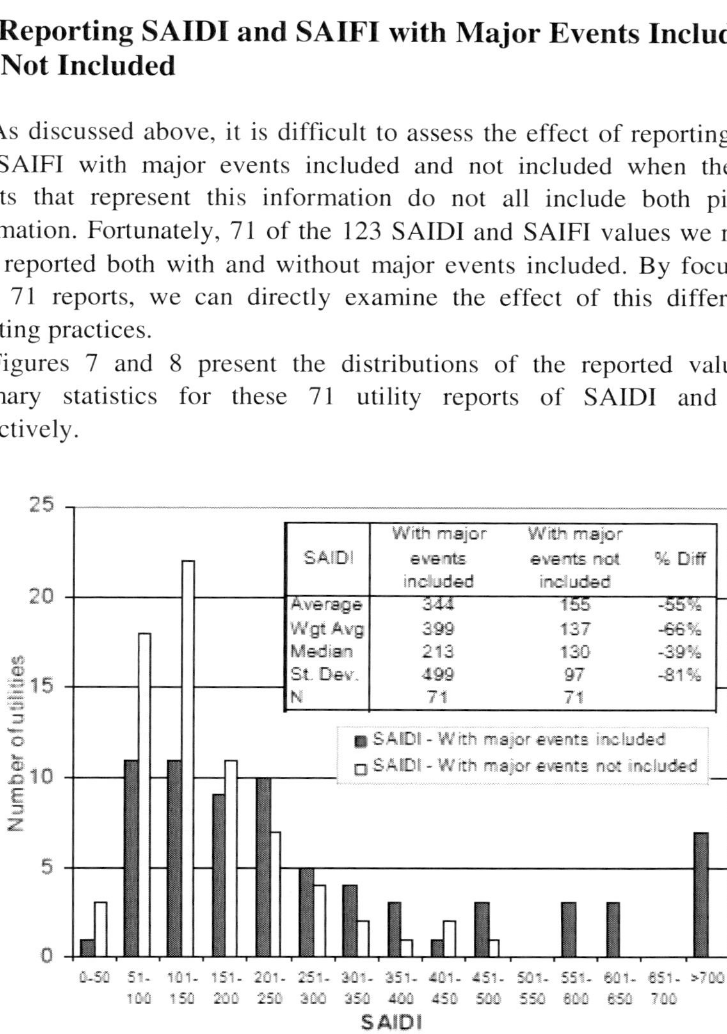

Figure 7. Distribution of and Summary Statistics for SAIDI Reported With Major Events Included and Not Included.

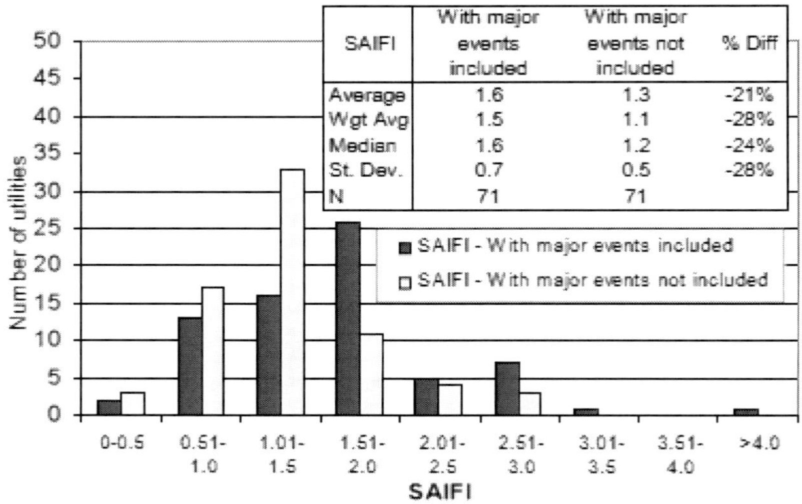

Figure 8. Distribution of and Summary Statistics for SAIFI Reported With Major Events Included and Not Included.

The effect of reporting with major events not included is clearly visible in the shift of the distributions for SAIDI and SAIFI when major events are included. The average SAIDI decreases by more than half (-55%) and the standard deviation is reduced by more than three quarters (-81%). The average, median, and standard deviation for SAIFI decrease by about one-quarter (-21%, -24%, and -28%, respectively).

4.6. Utility Practices for Defining Major Events

In reviewing how reporting with and without major events included affects SAIDI and SAIFI, we also find that utility practices vary regarding how major events are defined. As reported in Section 3, although IEEE Standard 1366-2003 provides a standardized, statistically based method for distinguishing among reliability events using the concept of Major Event Days (see Appendix A), this standard has only been formally adopted by four state PUCs. Fifteen state PUCs require use of a different definition for major events and an additional 16 state PUCs, which require routine reporting, have no formal requirements for the definition of major events.

With the exception of IEEE Standard 1366-2003, we find that the factors used to segment major events can be categorized according to the following characteristics:

- Whether the interruption was planned or unplanned. For example, an unplanned interruption might include one caused by a severe storm, earthquake, tornado, or high wind, while a planned interruption might refer to scheduled maintenance;
- The magnitude of the interruption, as measured by either the amount of load or number of customers affected; and
- The duration of the interruption, e.g., 24 hours or more.

Typically, more than one characteristic is used. Here are three examples drawn from actual definitions of major events in the utility-reported information we received for this study. In the first two examples, major events are called *excludable events.*

Example 1:

"An excludable event can be caused by: (a) planned service interruption, (b) a storm named by the National Hurricane Center, (c) a tornado recorded by the National Weather Service, (d) ice on lines, (e) a planned load management event, (f) any electric generation or transmission event not governed by subsections rules, or (g) an extreme weather or fire event causing activation of the county emergency operation center."

In this example, the major event definition includes both planned interruptions (planned service interruptions and planned load management events) and unplanned interruptions caused by various types of severe weather or fires.

Example 2:

"Excludable events are 'abnormal' situations such as hurricanes, tsunamis, earthquakes, floods, catastrophic equipment failures, and a single equipment outage that cascades into a loss of load that is greater than 10% of the system peak load."

In this example, the major event definition includes only unplanned interruptions ("abnormal" situations) but also considers the magnitude of the event, specifying a threshold for the amount of load that must be interrupted (10%).

Example 3:

> "A 'major event' is defined as any day (24-hour period) where 10% of the total company customer base experiences an interruption. The 'major event' ends when the total customers interrupted drops below 100."

In this example, the major event does not distinguish between planned and unplanned interruptions, but it does specify a load-based criterion (10% of customers).

Figure 9 summarizes our efforts to classify major event definitions using the above characteristics. Because more than one characteristic is typically used to define a major event, the sum of the factors used to define major events is greater than the number of utilities for which we received information. We find that the majority of major event definitions focus only on unplanned interruptions and rely on the number of customers interrupted to establish a threshold for the magnitude of the event. A large number of definitions also include a measure of the duration of the interruption.

Notably, 14 utility reports indicate that IEEE Standard 1366-2003 is used to segment the data using the concept of Major Event Days. We turn to a direct examination of the effect of reporting using this standard in the next sub-section.

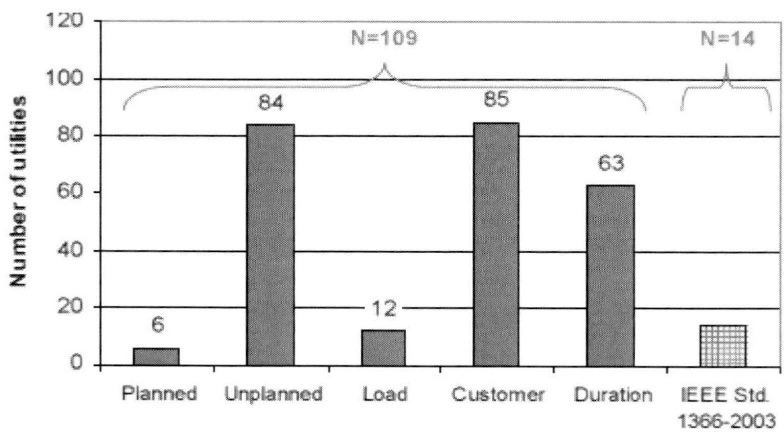

Figure 9. Assessment of Common Characteristics Used by Utilities to Define Major Events.

4.7. Using IEEE Standard 1366-2003 to Segment SAIDI and SAIFI with Major Event Days

IEEE Standard 1366-2003 is a voluntary industry effort to develop standard definitions and procedures for measuring and reporting reliability performance. Adoption of such a standardized approach would greatly simplify review of this information and improve assessment of utility-reported reliability information because it would eliminate differences in the reported information that are due solely to differences in reporting practices.

To date, however, adoption of the standard is in its infancy. As noted in Section 3, only four states formally require utility reporting using the standard and, among the 123 utilities for which we have information, 14 utilities reported reliability information using the standard.

Among these 14 utility reports, 9 give sufficient information to enable a preliminary assessment of the impact of using the new standard. That is, these 9 reported SAIDI and SAIFI calculated using both the concept of Major Event Days embodied in the standard as well using the former definition for major events to segment the reliability information.

Figure 10 shows the differences in the reported SAIDI and SAIFI that result from application of the two different procedures for segmenting the information. The differences are expressed as a percentage difference between the two values that result from use of each procedure.

From this preliminary examination, we find that there is no discernable bias introduced by use of the segmentation procedure embodied in the new IEEE standard compared to former practices for segmenting the information. That is, in several instances use of the procedure in the new standard leads to higher SAIDI and SAIFI values; in other instances it leads to lower values. Moreover, in this limited sample, the number of higher and lower values is roughly equal.

5. Comparison of Reliability Information Reported to State PUCs and to National Bodies

Both DOE and NERC require reporting of major electricity system incidents and disturbance events. Reporting to these national bodies is now mandatory, required in near-real time, and includes incidents that sometimes result in no loss of electric service to customers.[6]

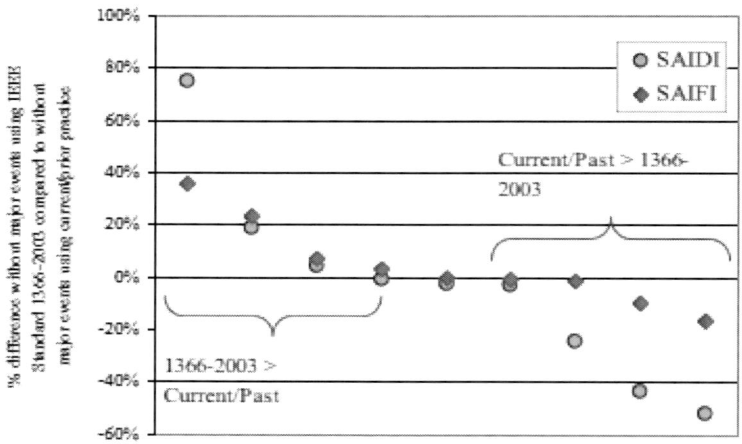

Figure 10. Comparison of SAIDI and SAIFI Without Major Events for 9 Utilities that Reported Using both IEEE Standard 1366-2003 and Current/Prior Practice for Segmenting Major Events.

The principle purpose of this form of reporting to national bodies is to provide information on major (i.e., very large) electricity system emergencies that may require immediate responses by industry or government to ensure public health and safety. Thus, there are fundamental and important differences between reports made to these national bodies and those made to state PUCs. With few exceptions, the same information is reported to both DOE and NERC at the same time.

Nevertheless, some of the incidents and disturbances reported to these national bodies are ones in which customers lose electric service. In this respect, despite fundamental differences in their purpose and scope, reported loss of electric service to these national bodies is similar to the utility-reported reliability information we collected from state PUCs. In this section, we report initial findings from our efforts to compare these two sources of information on power interruptions.

5.1. Department of Energy Form OE-417

DOE requires reporting of major electricity system incidents and disturbances. The information is reported via the Office of Electricity Delivery and Energy Reliability (OE) Form 417, commonly referred to as OE-417

(Department of Energy 2006). An initial notification is due to DOE within 60 minutes of the system disruption, followed by a detailed final report within 48 hours. The criteria for filing a report are as follows.

Form OE-417 must be submitted to the DOE Operations Center within one hour if any one of the following criteria is met:

1) Actual physical attack that causes major interruptions or impacts to critical infrastructure facilities or to operations;
2) Actual cyber or communications attack that causes major interruptions of electrical system operations;
3) Complete operational failure or shut-down of the transmission and/or distribution electrical system;
4) Electrical System Separation (islanding) where part(s) of a power grid remains operational in an otherwise blacked out area or within the partial failure of an integrated electrical system;
5) Uncontrolled loss of 300 Megawatts (MW) or more of firm system loads for more than 15 minutes from a single incident;
6) Load shedding of 100 MW or more implemented under emergency operational policy
7) System-wide voltage reductions of 3 percent or more; or
8) Public appeal to reduce the use of electricity for purposes of maintaining the continuity of the electric power system.

Form OE-417 must be submitted to the DOE Operations Center within six hours if one of the following applies and none of the eight categories above apply:

9) Suspected physical attacks that could impact electric power system adequacy or reliability or vandalism which targets components of any security systems;
10) Suspected cyber or communications attacks that could impact electric power system adequacy or vulnerability;
11) Loss of electric service to more than 50,000 customers for one hour or more; or
12) Fuel supply emergencies that could impact electric power system adequacy or reliability.

The information received by DOE on Schedule 1 is publicly available. The information collected on OE-417 includes the number of customers affected, demand involved, type and cause of incident, and the date and time the incident began and ended. For this analysis, we downloaded the information on the 91 events reported in year 2006 that involved customer interruptions.

5.2. North American Electric Reliability Corporation Disturbance Analysis Working Group Database

NERC standards, which became mandatory in June 2007, require reporting of major electricity system emergencies (North American Electric Reliability Corporation (NERC) 2006). When utilities send OE-417 to DOE, they also send the form to the Regional Reliability Organization and NERC under Standard EOP-004-1.

The requirements for submitting a report to NERC are contained in Standard EOP-004-1, Attachment 1 as follows:

1. Loss of a bulk power transmission component that significantly affects the integrity of the interconnection system operation. Generally, a disturbance report will be required if the event results in actions such as:
   a. Modification of operating procedures.
   b. Modification of equipment (e.g. control systems or special protection systems) to prevent reoccurrence of the event.
   c. Identification of valuable lessons learned.
   d. Identification of non-compliance with NERC standards or policies.
   e. Identification of a disturbance that is beyond recognized criteria, i.e. three-phase fault with breaker failure, etc.
   f. Frequency or voltage going below the under-frequency or under-voltage load shed points.
2. Occurrence of an interconnected system separation or system islanding or both.
3. Loss of generation by a Generator Operator, Balancing Authority, or Load-Serving Entity – 2,000 MW or more in the Eastern or Western Interconnection and 1,000 MW or more in the Electric Reliability Council of Texas (ERCOT) Interconnection.
4. Equipment failures/system operational actions, which result in the loss of firm system demand for more than 15 minutes, as described below:

   a. Entities with a previous year recorded peak demand of more than 3,000 MW are required to report all such losses of firm demands totaling more than 300 MW.
   b. All other entities are required to report all losses of firm demand totaling more than 200 MW or 50% of the total number of customers being supplied immediately prior to the incident, whichever is less.
5. Firm load shedding of 100 MW or more to maintain the continuity of the bulk electric system.
6. Any action taken by a Generator Operator, Transmission Operator, Balancing Authority, or Load-Serving Entity that results in:
   a. Sustained voltage excursions equal to or greater than 10%, or
   b. Major damage to power system components, or
   c. Failure, degradation, or misoperation of system protection, special protection schemes, remedial action schemes, or other operating systems that do not require system operator intervention, which did result in, or could have resulted in, a system disturbance as defined by steps 1 through 5 above.
7. An Interconnection Reliability Operating Limit violation as required in reliability standard TOP-007.
8. Any event that the Operating Committee requests to be submitted to Disturbance Analysis Working Group (DAWG) for review because of the nature of the disturbance and the insight and lessons the electricity supply and delivery industry could learn.

NERC compiles the filed information in a database that is maintained by DAWG. The DAWG database contains a summary of the disturbance, including the date, region, associated utilities, type of event, MW lost, number of customers affected, and cause. The information contained in the DAWG database is similar to information reported on OE-417, but contains additional information on the disturbance and actions taken.

The information in the DAWG database is publicly available. We obtained information for a total of 52 events in 2006 involving incidents where customers were affected. Prior to June 2007, filing information with NERC on major system emergencies was voluntary.

Consequently, there may be more total events listed in OE-417 than are listed in the DAWG database for 2006.[7]

5.3. Comparison of Reported Reliability Information

We compared the interruptions reported to state PUCs and to national bodies to examine the relationship between the reliability information reported by utilities to each of these entities. Reliability information reported to national bodies is restricted to significant events involving large numbers of customers (as defined by the reporting criteria that trigger the filing requirement). Reliability information reported to state PUCs is inclusive and, in principle, includes any and all events affecting customers, regardless of the number of customers affected.[8]

Comparing the interruptions reported to national bodies, in aggregate, to the total interruptions recorded by utilities, in aggregate, gives a rough measure of the significance of large interruptions compared to all interruptions.

For this comparison, we calculated a "bulk power system only" SAIDI and SAIFI value from the information contained in the utility reports to DOE and NERC. Calculating a pseudo value like this means we are effectively treating the entire U.S. as a single utility, with each report documenting a single, national-level event recorded on the transmission system.[9]

Expressing utility reliability information reported to DOE and NERC using the definitions for SAIDI and SAIFI, however, requires important additional assumptions beyond that reported to these national bodies. The necessary assumptions we made are reasonable, but they leave ample room for improvement. Hence, the values we calculate should be regarded as very preliminary and rough estimates.

To calculate SAIFI from information reported to DOE and NERC required two assumptions. First, we assumed that each reported event was a distinct event from that which might be reported by another entity. This assumption is incorrect when two or more entities report interruptions stemming from the same initiating cause (e.g., a widespread weather event, such as a major hurricane). Second, we assumed that each reported event represented a single event. This assumption is incorrect when an entity aggregates multiple interruptions over a period of time into a single report (e.g., when multiple interruptions result from multi-day period severe weather events).

To calculate SAIDI from the information reported to DOE and NERC required an assumption on how the reported number of customers affected changed over the reported duration of the interruption. We assumed that all the reported number of customers affected would initially experience the interruption and that the number would decline exponentially over the duration of an interruption. This assumption is reasonable for interruptions that are initiated by a single triggering event followed by a period of sequential restoration of service. This assumption is not appropriate, however, for interruptions that represent multiple interruptions that affect different numbers of customers at different periods of time, such as interruptions resulting from a multi-day period of severe weather.

With these important limiting assumptions in mind, Table 5 reports the SAIDI and SAIFI values we calculated from the information reported to DOE and NERC and compared them to the average SAIDI and SAIFI with major events included reported from our data collected from the PUCs in Table 4. This comparison is intended to provide a first-order estimate of the portion of interruptions experienced by utilities that result from events large enough to be reported to national bodies.

We find that large interruptions appear to represent a small portion of the total number of interruptions experienced by customers (as shown by SAIFI in Table 5). The SAIDI value calculated from the information reported to national bodies suggests that events large enough to warrant reporting to national bodies accounts for a significant amount of the interruptions experienced by utility customers. For example, the SAIDI value calculated from the information reported to DOE suggests that more than half of the customers affected by the total duration of interruptions are accounted for by the large events that must be reported to DOE.

Table 5. Comparison of Year 2006 Reliability Information on Major Electricity System Events to Reliability Information on All Events

	SAIFI	SAIDI
LBNL estimate derived from DOE OE-417	0.10	164
LBNL estimate derived from NERC DAWG	0.07	110
Average of 78 utilities whose reports included major events	1.54	292

These findings, however, are preliminary. They depend on a number of important assumptions that cannot be resolved solely on the basis of information that has been reported to these national bodies. Moreover, in order to provide a reasonable basis for this comparison, we have focused on the 80 utilities that report SAIDI and SAIFI with major events included. Obviously, based on these reports alone, we cannot generalize that the reliability information reported by these 80 utilities are representative of the other 43 utilities we received information on, or of the utilities representing roughly 40% of U.S. electricity sales we have no reliability information.

Nevertheless, to a first approximation, this comparison does support the contention, which is widespread within the utility industry, that the majority of interruptions experienced by customers, at least in number, are local in nature.

5.4. Consistency in Reporting of Large Power Interruptions

In reviewing information regarding major events that was reported by utilities to state PUCs, we identified a second means for comparing this reliability information with what is reported to national bodies.

A large number of utilities (55 of the 123) report information on each individual major event along with SAIDI and SAIFI calculated with major events included or not included. The information reported on each major event to state PUCs is similar to the information reported to national bodies in that it includes the date, time, location, and sometimes the number of customers affected and duration of the event.

Because events that might qualify as major events for the purpose of isolating an individual utility's SAIDI and SAIFI are the largest events experienced by that utility, we expected that some of these events would also qualify for reporting the national bodies. Thus, we attempted to match the information on individual major events reported to state PUCs to the information on individual events reported to national bodies. We first sought to match events according to the reporting utility, date, and time of the interruption. We then sought to compare the number of customers affected. Although this comparison is incomplete, it provides preliminary insight into the consistency of information reported to state PUCs and these national bodies.

For a number of reasons, we expected this matching to yield low results. First, as noted in footnote 7, some utilities exclude events reported to national bodies from major events that are listed individually in reports to state PUCs. Second, reports to national bodies are made by utilities that do not report reliability information to state PUCs (either because utilities are not state regulated, or the state does not have a reporting requirement). Third, the events that meet criteria for reporting to national bodies may not meet the criteria for treatment as major events for state PUC reporting purposes. Fourth, we do not have a census of major events that are recorded by all utilities that also provide reliability information to state PUCs; we have lists of major events for only 55 of the 123 such utilities.

In matching the year 2006 events reported to DOE on Form OE-417 to events reported to state PUCs, we found matches for 24 events based on date, time, and location. Three of these matches were dropped from further review because the entity reporting to OE was not the same entity that reported to the state PUC. An additional 13 matches were dropped from further review because the information reported to state the PUC did not include the number of affected customers.

The percentage difference in the number of affected customers reported to DOE compared to that reported to the state PUC for the remaining eight events is shown in Figure 11. A negative percentage means that the number of affected customers reported to the state PUC is less than the number reported to DOE; a positive percentage means that the number reported to the state PUC is greater than that reported to DOE.

Figure 11 shows that, for this small number of matched events, there are sometimes very large differences. Reports for more than half of the events differ by more than 20%. For the same event, reports to state PUCs appear to indicate more customers are affected compared to reports to national bodies.

In matching events reported to NERC in 2006 with those reported to PUCs, we found matches for 16 of these events based on date, time, and location. Six of these matches were dropped from further review because the entity reporting to NERC was not the same entity that reported to the state PUC. An additional two matches were dropped from further review because the information reported to state PUCs did not include the number of affected customers. Interestingly, the remaining eight matched events using the NERC DAWG information were not the same events matched previously using the DOE OE-417 information.

The percentage difference in the number of affected customers reported to NERC compared to the number reported to the state PUC for the remaining eight events are shown in Figure 12. A negative percentage means that the number of affected customers reported to the state PUC is less than the number reported to NERC; a positive percentage means that the number reported to the state PUC is greater than that reported to NERC.

Figure 12 shows, as did Figure 11, some very large differences. Reports for more than half of the events differ by more than 20%. Here, the tendency of state PUCs to indicate more customers are affected compared to reports to national bodies is even greater than in the previous comparison.

It is difficult to draw definitive conclusions from this initial assessment of reliability information reported to national bodies compared to that reported to state PUCs. Some of the reasons we expected to find such few matches have already been listed. A final consideration is that reporting to national bodies occurs in near real-time. Both DOE and NERC require immediate notification and preliminary information within 24 hours of an event, with a final filing within 48 hours for DOE and within 60 days for NERC. Reliability information reported to state PUCs typically takes the form of an annual summary, provided sometime after the end of a given reporting period.

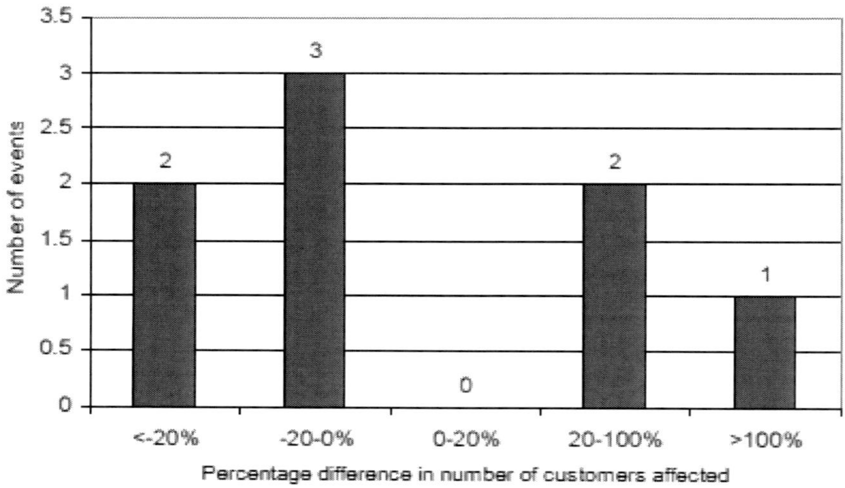

Figure 11. The Difference Between the Number of Customers Affected, as Reported on OE Form 417, and as Reported to State PUCs for Eight Major Events in 2006.

6. Summary of Findings and Conclusions

Our findings regarding state PUC practices and rules on reliability information reported by utilities are summarized as follows:

- Thirty-five state PUCs, including DC, require routine reporting of reliability event information. This is a net increase of 10 state PUCs over the number reported in a similar survey conducted by the NRRI in 2004.

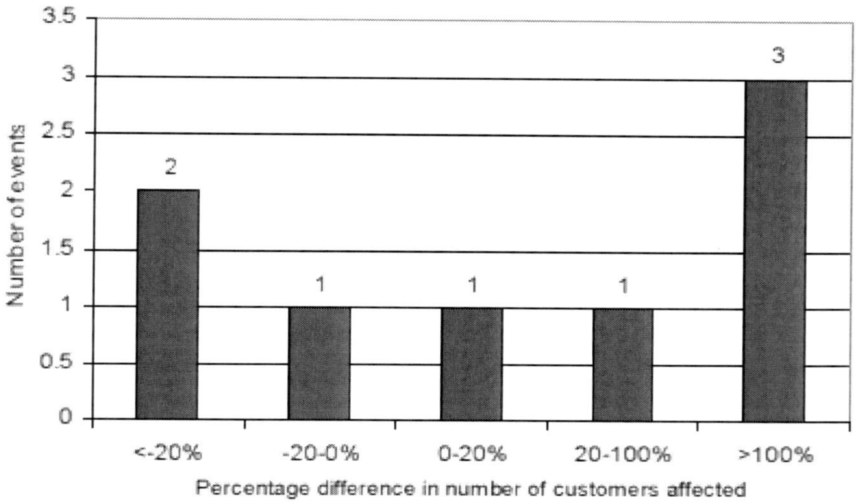

Figure 12. The Difference Between the Number of Customers Affected, as Reported to NERC, and as Reported to State PUCs for Eight Major Events in 2006.

- These 35 PUCs require annual reporting of SAIDI and SAIFI and/or CAIDI, which, along with SAIFI, can be used to derive SAIDI. Only two state PUCs require reporting of MAIFI.
- Twenty-one PUCs have reporting requirements that formally define major events. Of these 21, four require reporting following the IEEE Standard 1366-2003, IEEE Guide for Electric Power Distribution Reliability Indices, which introduces a consistent means for defining major events using the concept of "major event days."
- An additional four PUCs receive reliability information from utilities, though not as a result of a formal reporting requirement.
- Thirty-seven state PUCs, including DC, make publicly available or summarize in publicly available documents the reliability information they collect from utilities.

Our findings regarding utility practices for collecting and reporting reliability information to state PUCs are summarized as follows:

- All utilities reported SAIDI and SAIFI (and/or CAIDI). Only 12 of the 123 utilities reported MAIFI.

- Summary statistics for reported SAIDI, SAIFI, and MAIFI exhibit observable though not statistically significant variations across census regions.
- The definition of and practices for recording sustained and momentary interruptions have evolved over time leading to inconsistencies among utilities.
- Differences in the definition of a sustained interruption do not appear to affect SAIDI or SAIFI in a statistically significant manner.
- Utilities define major events as a means for distinguishing between utility performance in planning for and responding to routine interruptions versus that for non-routine or extraordinary interruptions.
- The definition of a major event is not consistent among the majority of utilities.
- IEEE Standard 1366-2003 introduces a consistent means for defining major events using the concept of "major event days."
- Some utilities report SAIDI and SAIFI both including and not including major events; other utilities only report SAIDI and SAIFI not including major events.
- When major events are not included, SAIDI is lowered relatively more than SAIFI compared to when major events are included.
- Many utilities report descriptive information on each major event
- Use of IEEE Standard 1366-2003 does not appear to bias SAIDI or SAIFI values compared to using prior definitions of major events.

We also collected information on bulk power system emergencies reported by utilities in near real-time to national bodies in 2006, including DOE and NERC, and compared aspects of this information to that reported by utilities to state PUCs. Our findings are summarized as follows:

- Information on electricity reliability reported to these two national bodies consists of descriptive information that is reported in near real-time on individual, large events that affect the bulk power system. The reporting takes place in near real-time because an important purpose of the reporting is to notify relevant industry and public bodies of significant power system events that may require immediate response. With few exceptions, the same information is reported to both DOE and NERC at the same time.

- Many, but not all, events reported to these national bodies also cause power interruptions to customers. For these events, the number of customers affected is reported.
- An initial assessment of these events supports the conventional wisdom that the majority of power interruptions experienced by customers are not due to large events that affect the bulk power system; they are due to more localized events that affect only utility distribution systems.
- It is difficult to cross-reference information reported to national bodies on individual large bulk power system events that cause power interruptions, as defined by these national bodies, with information reported to state PUCs on individual major events, as defined by either the PUC or the reporting utility.

From these findings, we draw the following conclusions and recommendations:

- State PUC interest in electricity reliability is growing.
- However, differences in utility reporting practices hamper meaningful comparisons of reliability information reported by utilities to different state PUCs and, therefore, may limit the effectiveness of efforts to measure the effectiveness of efforts to improve reliability.
- Efforts to eliminate differences that are solely due to reporting practices are just beginning. These efforts, which focus on using standard definitions, such as those promoted by IEEE Standard 1366-2003, are promising and should be encouraged.
- Until IEEE Standard 1366-2003 is adopted universally, regulators concerned about the definition and treatment of major events in reporting reliability information should consider requiring reporting of SAIDI and SAIFI both including and not including major events, as well as descriptive information on each major event.
- More work is required to better understand the sources of discrepancies and the importance of seeking greater consistency between reliability information reported to national bodies and that reported to state PUCs.

APPENDIX A. IEEE STANDARD 1366-2003 RELIABILITY INDEX AND MAJOR EVENT DAY DEFINITIONS

Utilities record, organize, and report information on reliability events using a variety of metrics. The System Average Interruption Frequency Index (SAIFI) and System Average Interruption Duration Index (SAIDI) measure the number and duration, respectively, of sustained interruptions experienced in one year by a customer (IEEE Working Group on System Design 1999). SAIDI and SAIFI are the two metrics most commonly used to express the reliability of electricity service based on duration and frequency of electricity interruptions. The Customer Average Interruption Duration Index (CAIDI) is also commonly reported; it is derived from SAIDI and SAIFI. The Momentary Average Interruption Frequency Index (MAIFI) is less commonly reported. However, MAIFI is important because SAIDI, SAIFI, and MAIFI, taken together, provide a comprehensive measure of electricity reliability, as measured by the duration and frequency of power interruptions.

According to the IEEE Standard 1366-2003 standard, the definitions for these indices are as follows:

$$\text{SAIDI} = \frac{\sum \text{Customer Interruption Durations}}{\text{Total Number of Customers Served}}$$

$$\text{SAIFI} = \frac{\sum \text{Total Number of Customers Interrupted}}{\text{Total Number of Customers Served}}$$

$$\text{CAIDI} = \frac{\text{SAIDI}}{\text{SAIFI}}$$

$$\text{MAIFI} = \frac{\sum \text{Total Number of Customer Momentary Interruptions}}{\text{Total Number of Customers Served}}$$

IEEE Standard 1366-2003 defines sustained interruptions as lasting more than five minutes and momentary interruptions as lasting five minutes or less.

IEEE Standard 1366-2003 also provides guidelines for identifying major events using the concepts of Major Event Days (MEDs). This standard introduces a “2.5 beta method” that defines an MED as a day in which the daily system SAIDI exceeds a threshold value, TMED. The TMED threshold value is calculated as follows:

a) Collect values of daily SAIDI for five sequential years ending on the last day of the last complete reporting period. If fewer than five years of historical data are available, use all available historical data until five years of historical data are available.
b) Only those days that have a SAIDI/Day value will be used to calculate the TMED (do not include days that did not have any interruptions).
c) Take the natural logarithm (ln) of each daily SAIDI value in the data set
d) Find a (Alpha), the average of the logarithms (also known as the log-average) of the data set.
e) Find /3 (Beta), the standard deviation of the logarithms (also known as the log-standard deviation) of the data set
f) Compute the MED TMED, using the following equation,

$$T_{MED} = e^{(\alpha + 2.5\beta)}$$

g) Any day with daily SAIDI greater than the threshold value TMED that occurs during the subsequent reporting period is classified as a MED.

REFERENCES

Department of Energy. 2004. *The August 14, 2003 Blackout One Year Later: Actions Taken in the United States and Canada To Reduce Blackout Risk*. Washington, DC: August 13. 18.

Department of Energy. 2006. "Electric Disturbance Events (OE-417)." Office of Electricity Delivery and Energy Reliability (OE). *http://www.oe.netl.doe.gov/oe417.aspx.*

Energy Information Administration. 2003. *The National Energy Modeling System: An Overview 2003*. Washington D.C.: EIA/DOE. March 2003. 72.

Energy Information Administration. 2006. "Form EIA-861 Final Data File for 2006." DOE/EIA. *http://www.eia.doe.gov/cneaf/electricity/page/eia861.html.*

Gonzalez, M. 2006. "IEEE Benchmarking Update."*http://grouper.ieee.org/groups/td/dist/sd/doc/2005-05-UpdateToEEIonIEEE1366.pdf*. May 20.

Hines, P., J. Apt, and S. Talukdar. 2008. "Trends in the History of Large Blackouts in the United States." *2008 IEEE Power and Energy Society (PES) General Meeting*, Pittsburgh, Pennsylvania, (20-24): 8.

IEEE Power Engineering Society. 2004. *IEEE Std 1366:-2003 IEEE Guide for Electric Power Distrubtion Reliability Indices*. ISBN 0-7381-3890-8 SS95193. New York: Institute of Electrical and Electronics Engineers, Inc. May 14. 35 pages.

IEEE Working Group on System Design. 1999. *1366-1998 Trial-Use Guide for Power Distribution Reliability Indices*. ISBN# 0738115479. New York: IEEE.

LaCommare, K. H., and J. H. Eto. 2004. *Understanding the Cost of Power Interruptions to U.S. Electricity Customers*. LBNL-55718. Berkeley: Lawrence Berkeley National Laboratory. September. 50 pages.

North American Electric Reliability Corporation (NERC). 2006. "DAWG Database: Disturbances, Load Reductions, and Unusual Occurrences." *http://www.nerc.com/files/disturb06.pdf.*

Warren, C. 2006. "IEEE 1366 & Regulatory Implications." Institute of Electric and Electronics Engineers, Inc. *http://grouper.ieee.org/groups/td/dist/sd/doc/2006-07-IEEE1366-RegulatoryImplications.pdf*. May 20.

End Notes

[1] For technical information on the definitions and calculation of SAIDI, SAIFI, and MAIFI, please see Appendix A.

[2] Note: There are no state-regulated electric utilities in Nebraska; thus, this state was not contacted.

[3] As noted previously, there are no state-regulated electric utilities in Nebraska; thus, this state was not contacted.

[4] Although a state PUC may not require routine reporting, some of these PUCs, in fact, receive reliability performance information from utilities and in some cases made this information available for this study, including Arkansas, Idaho, North Dakota, and New Mexico.

[5] This discussion has been informed by comments received from members of the IEEE Distribution Reliability Working Group on an early draft of this chapter.

[6] Reporting to NERC became mandatory in June, 2007. Prior to June 2007, including the period analyzed in this section, reporting to NERC was encouraged, but voluntary.

[7] Utility practices vary regarding the treatment of interruptions, which are large enough to warrant immediate reporting to national bodies, with respect to inclusion of these interruptions in the routine reports to PUCs, which we review in this chapter. Most utilities include information on all interruptions in their calculation of SAIDI and SAIFI. Some utilities, however, appear to exclude information on these large interruptions from their calculation of SAIDI and SAIFI.

[8] As noted in the previous footnote, if utility reporting practices exclude events large enough to warrant immediate reporting to national bodies from their SAIDI and SAIFI values, these "bulk-power system" SAIDI and SAIFI values should be considered additive to the local utility SAIDI and SAIFI values. If utility reporting practices include all interruptions, including these large interruptions, then these "bulk-power system" SAIDI and SAIFI values should be considered a portion of (or contributor to) the local utility SAIDI and SAIFI values.

[9] See Hines et al. 2008 for more additional discussion of this approach (Hines et al. 2008).

In: Electricity Reliability in the United States ISBN: 978-1-62417-149-9
Editor: Emmanuel Martinez

Chapter 4

WEATHER-RELATED POWER OUTAGES AND ELECTRIC SYSTEM RESILIENCY*

Richard J. Campbell

SUMMARY

High winds, especially when combined with precipitation from seasonal storms, can cause damage to electricity utility systems, resulting in service interruptions to large numbers of electricity customers. While most such power outages are caused by damage from trees and tree limbs falling on local electricity distribution lines and poles, major power outages tend to be caused by damage to electricity transmission lines which carry bulk power long distances. Depending on the severity of the storm and resulting impairment, power outages can last a few hours or extend to periods of several days, and have real economic effects. Power outages can impact businesses (primarily through lost orders and damage to perishable goods and inventories), and manufacturers (mainly through downtime and lost production, or equipment damage). Data from various studies lead to cost estimates from storm-related outages to the U.S. economy at between \$20 billion and \$55 billion annually. Data also suggest the trend of outages from weather-related events is increasing.

* This is an edited, reformatted and augmented version of a Congressional Research Service publication, CRS Report for Congress R42696, prepared for Members and Committees of Congress, dated August 28, 2012.

Suggested solutions for reducing impacts from weather-related outages include improved tree-trimming schedules to keep rights-of-way clear, placing distribution and some transmission lines underground, implementing Smart Grid improvements to enhance power system operations and control, inclusion of more distributed generation, and changing utility maintenance practices and metrics to focus on power system reliability. However, most of these potential solutions come with high costs which must be balanced against the perceived benefits.

A number of options exist for Congress to consider which could help reduce storm-related outages. These range from improving the quality of data on storm-related outages, to a greater strategic investment in the U.S. electricity grid. Congress could empower a federal agency to develop standards for the consistent reporting of power outage data. While responsibility for the reliability of the bulk electric system is under the Federal Energy Regulatory Commission (as per the Energy Policy Act of 2005), no central responsibility exists for the reliability of distribution systems. One possible option could be to bring distribution systems under the Electric Reliability Organization for reliability purposes. Recovery after storm-related outages might be enhanced by a federal role in formalizing the review or coordination of electric utility mutual assistance agreements (MAAs). This would not necessarily mean federal approval of MAAs, but may help in the cooperative coordination of additional federal and state resources, especially in a wide, multi-state weather event. While there has been much discussion of transmission system inadequacies and inefficiencies, many distribution systems are in dire need of upgrades or repairs. The cost of upgrading the U.S. grid to meet future uses is expected to be high, with the American Society of Civil Engineers estimating a need of $673 billion by 2020. While the federal government recently made funding available of almost $16 billion for specific Smart Grid projects and new transmission lines under the American Recovery and Reinvestment Act of 2009, there has not been a comprehensive effort to study the needs, set goals, and provide targeted funding for modernization of the U.S. grid as part of a long-term national energy strategy. Such an effort would also require decisions about the appropriate roles of government and the private sector.

Power delivery systems are most vulnerable to storms and extreme weather events. Improving the overall condition and efficiency of the power delivery system can only serve to improve the resiliency of the system, and help hasten recovery from weather-related outages. Ultimately, however, electric utilities are responsible for this infrastructure. They are in the business of selling electricity, and they cannot sell electricity if their power delivery systems are out of service.

INTRODUCTION

High winds, especially when combined with precipitation from seasonal storms, can cause damage to electricity utility systems, resulting in service interruptions to large numbers of electricity customers. While most such power outages[1] are caused by damage from trees and tree limbs falling on local electricity distribution lines and poles, major power outages tend to be caused by damage to electricity transmission lines, which carry bulk power long distances.

Depending on the severity of the storm and resulting impairment, power outages can last a few hours or extend to periods of several days. This in turn can have real economic effects, as power outages can impact businesses (primarily through lost orders and damage to perishable goods and inventories), and manufacturers (mainly through downtime and lost production, or equipment damage).

Potential issues for Congress concern the resiliency of the electric power system, and whether the economic impacts of storm-related power outages require additional measures to be undertaken by the federal government in mitigating storm-related outages. Congress has already recognized the importance of the reliable operation of the bulk power system with the Energy Policy Act of 2005[2] (EPACT) authorization of an Electric Reliability Organization[3] (ERO).

This chapter will focus on the impacts of sustained power outages as might result from the result of seasonal storms, and whether there is a role for the federal government in hastening the restoration of power from weather-related outages.

ANATOMY OF WEATHER-RELATED POWER OUTAGES

Electric power is generated and sent over transmission lines to substations which reduce the voltage levels for distribution to end-use customers, as shown in Figure 1. The network of cables enabling electric power to be sent to customers generally exists in an exterior or "above ground" environment largely exposed to the elements. As such, power outages can result from seasonal storms which often combine the furies of wind, rain, snow, or ice. The more severe weather events usually cause the greatest damage to electric power transmission and distribution infrastructure as damage can result from trees or branches falling on electricity lines. While data on storm-related power outages exist, they are not generally considered to be complete or well-characterized with regard to the cause of the outage event. It has been estimated that 90% of customer outage-minutes are due to events which affect local distribution systems.4 However, the remaining 10% stem from generation and transmission problems, which can cause wider-scale outages affecting larger numbers of customers.

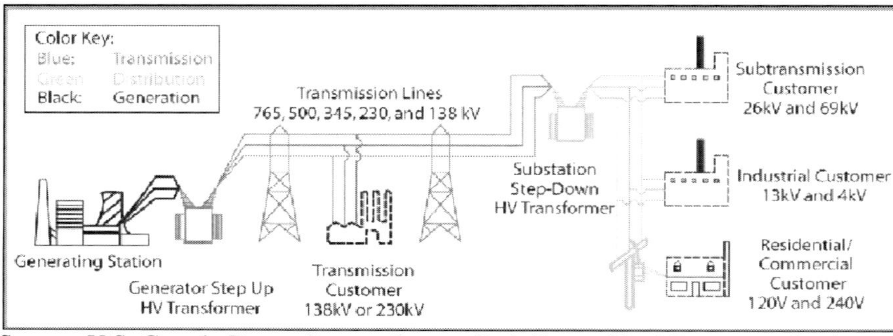

Source: U.S.-Canada Power System Outage Task Force, Final Report on the August 14, 2003, Blackout in the United States and Canada: Causes and Recommendations, April 2004, p. 5, https://reports.energy.gov/BlackoutFinal-Web.pdf.

Figure 1. Electric Power System Elements.

U.S. Storm-Related Outages

The North American Electric Reliability Corporation (NERC) requires electric utilities to report events which cause disturbances that interrupt service (i.e., power outages) of more than 300 MegaWatts (MW) or affect 50,000 customers or more. A University of Vermont analysis5 of NERC data describes 933 events causing outages from the years 1984 to 2006, and is presented in Table 1.

According to the Vermont study, almost 44% of the events in the period were weather-related (i.e., caused by tornado, hurricane/tropical storm, ice storm, lightning, wind/rain, or other cold weather). The study noted that the data include many events smaller than the NERC reporting threshold. It also noted that some of the reported events have "multiple initiating" causes, since some events (such as lightning) can trigger other outages or operator errors.

Table 1. Large Blackouts in the United States

Statistics for Outage Cause Categories

	% of events	Mean size in MW	Mean size in customers
Earthquake	0.8	1,408	375,900
Tornado	2.8	367	115,439
Hurricane/Tropical Storm	4.2	1,309	782,695
Ice Storm	5	1,152	343,448
Lightning	11.3	270	70,944
Wind/Rain	14.8	793	185,199
Other cold weather	5.5	542	150,255
Fire	5.2	431	111,244
Intentional attack	1.6	340	24,572
Supply shortage	5.3	341	138,957
Other external cause	4.8	710	246,071
Equipment Failure	29.7	379	57,140
Operator Error	10.1	489	105,322
Voltage reduction	7.7	153	212,900
Volunteer reduction	5.9	190	134,543

Source: Trends in the History of Large Blackouts in the United States, http://www.uvm.edu/~phines/publications/ 2008/Hines_2008_blackouts.pdf.

Notes: Totals are greater than 100% because some events fall into multiple initiating-event categories.

A 2004 study[6] by Lawrence Berkeley Laboratory (LBL) looking at power interruptions characterized power outages as being of *short* duration lasting less than five minutes, and *sustained* duration outages[7] lasting longer than five minutes (and extending to hours or days). Power outages caused by storm-related events can vary in duration but tend to be sustained disruptions. The study noted that weather-related events are not always captured in power outage data.[8]

The U.S. Department of Energy (DOE) maintains its own database of grid disturbance events.[9]

A recent analysis[10] by LBL's Evan Mills of the DOE database shows an increasing number of outages from 1992 to 2010, all dominated by weather-related events (see *Figure 2*). According to Mills, approximately 78% of the reported 1,333 electric grid disruptions in the period were weather-related (i.e., caused by temperature extremes, ice/snow/winter storm, thunderstorm/tornado/lightning, windstorm/hurricane/severe storm, and undefined weather events), and the grid disruptions affected 178 million metered customers.

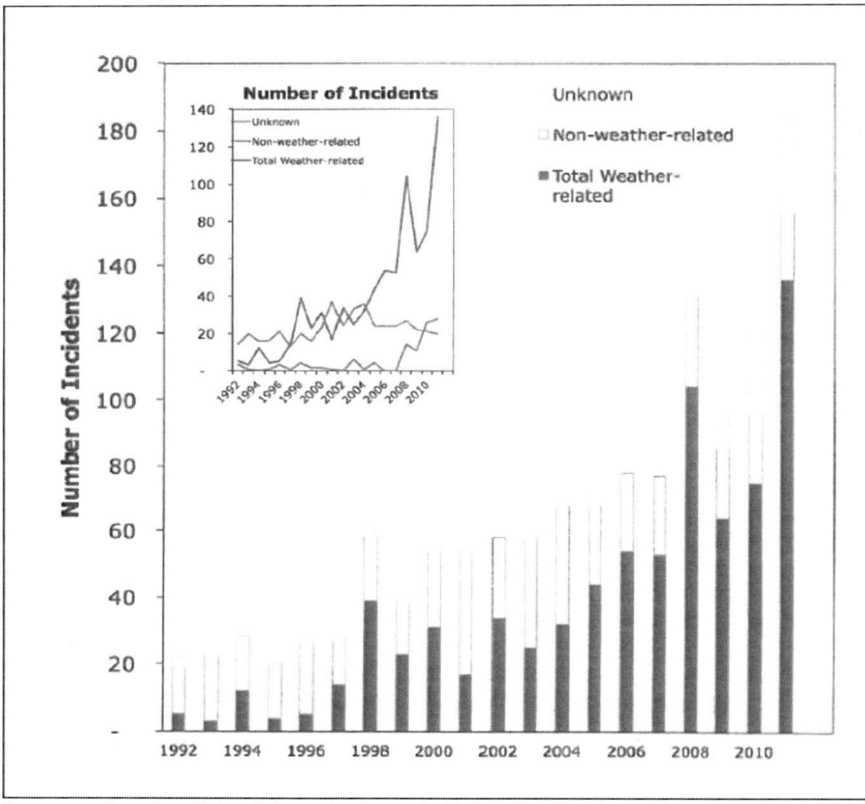

Source: Electric Grid Disruptions and Extreme Weather. See http://evanmills.lbl.gov/presentations/Mills-GridDisruptions-NCDC-3May2012.pdf.

Notes: Historical "Grid Disturbance" data from the U.S. Department of Energy, Energy Information Administration. Form OE-417, "Electric Emergency Incident and Disturbance Report" (and before 1978 from the National Electric Reliability Council, Disturbance Analysis Working Group).

Figure 2. Significant U.S Grid Weather-Related Grid Disturbances. (With Inset of Non-Weather- vs. Weather-Related Outage Comparison)

According to graphs shown in *Figure 2*, the number of observed outages from weather-related incidents seems to be increasing. Mills believes the reasons for the increased trend in outages may be due to a combination of power grid deterioration and a real increase in the number of observed extreme weather events.[11]

Outage Rates in Other Countries

The United States is generally considered to have one of the industrial world's most reliable electric power systems. However, when compared statistically to other nations, the U.S. grid does not necessarily meet those expectations.

There are two main indices generally used to measure reliability. The *system average interruption duration index*[12] (SAIDI) represents the average amount of time per year that power supply to a customer is interrupted, expressed in minutes per customer per year. The *system average interruption frequency index*[13] (SAIFI) represents the average number of times per year that the supply to a customer is interrupted, expressed as interruptions per customer per year. However, there is a lack of consistency in how the inputs to these indices are measured, both domestically and internationally. Much of the discrepancy again concerns whether or how storm-related outage events are counted as outage events. Some jurisdictions, both in the United States and internationally, consider storm-related outages as "extreme" events, and thus are not included in power outage statistics. Additionally, what is considered as unusual weather in one region may not be counted as unusual in another region.

> It should be noted, however, that weather circumstances that occur occasionally should not be considered as exceptional events. For example, snowstorms are not an exceptional event in Sweden, but could be seen as an exceptional event in southern Greece. Similarly, very hot temperature for sustained periods of time is not an exceptional event in Greece, but could be considered so in Sweden. Lightning should not be treated as an exceptional event anywhere in Europe.[14]

SAIDI and SAIFI reliability indices for nine industrial countries are summarized in Table 2. As can be seen, the United States has the highest average annual outage time per customer, and the third-highest average annual number of supply outages per customer.

It should be noted that many European countries consider the threshold for defining an extended outage as longer than three minutes[15] instead of the five minute standard generally used in the United States. European countries also commonly have a single regulatory body which is responsible for reliability of the entire nation's power system.

These authorities generally authorize expenditures for improving the reliability of service.

Table 2. Comparison of International Reliability Indices

Country	SAIDI	SAIFI
United States	240	1.5
Austria	72	0.9
Denmark	24	0.5
France	62	1.0
Germany	23	0.5
Italy	58	2.2
Netherlands	33	0.3
Spain	104	2.2
United Kingdom	90	0.8

Source: Galvin Electricity Initiative, Electric Reliability: Problems, Progress and Policy Solutions. See http://www.galvinpower.org/sites/default/files/Electricity_Reliability_031611.pdf.

Overseeing the reliability of the bulk power system in the United States is the responsibility of the Federal Energy Regulatory Commission (FERC) as per EPACT, but responsibility for maintaining the reliability of the nation's electricity distribution systems is the responsibility of many state and local regulatory bodies. It is also likely that the physical size of the United States, and the historical weather extremes possible in any given season across the country, contribute to the U.S. standing in the statistics of *Table 2*.

However, the generally degraded condition of the U.S. electric grid's infrastructure also likely contributes to the statistical quandary. The American Society of Civil Engineers (ASCE) has labeled the nation's grid as a "patchwork system" that may ultimately break down without a massive investment of an estimated $673 billion by 2020.[16] ASCE estimated that over the next decade, there is a gap between what it viewed as needed and current spending annually of $12.3 billion for power generation, $37.3 billion for transmission systems, and $57.4 billion for distribution systems.[17]

Economic Costs Associated with Storm-Related Power Outages

Power outages can impact electricity consumers primarily through property loss and business disruption. This can result in lost orders, and damage to perishable goods and inventories for businesses. Power outages can critically affect manufacturing operations mainly through downtime as workers are idled, and potentially damage equipment and production processes. The loss of electricity can also disrupt many other electric power-dependent activities sometimes taken for granted (such as traffic control) which can result in further economic impacts.

A study in 2001 by the Electric Power Research Institute (EPRI) estimated the costs to the economy of power disturbances.[18] The study found that three sectors of the economy were particularly vulnerable to power outages:

> The digital economy (DE). This sector includes firms that rely heavily on data storage and retrieval, data processing, or research and development operations. Specific industries include telecommunications, data storage and retrieval services (including collocation facilities or Internet hotels), biotechnology, electronics manufacturing, and the financial industry.
> Continuous process manufacturing (CPM). This sector includes manufacturing facilities that continuously feed raw materials, often at high temperatures, through an industrial process. Specific industries include paper; chemicals; petroleum; rubber and plastic; stone, clay, and glass; and primary metals.
> Fabrication and essential services (FES). This sector includes all other manufacturing industries, plus utilities and transportation facilities such as railroads and mass transit, water and wastewater treatment, and gas utilities and pipelines.

These three sectors accounted for almost 40% of U.S. gross domestic product[19] in 2001, but disruptions to especially the DE and FES sectors impact other sectors of the economy through the services they provide.

EPRI recognized that the economic cost of power outages was largely related to the length of the outage (while noting even short duration outages of a few minutes could have large costs), and estimated the average cost of a one-hour outage for manufacturing and DE firms at $7,795 per firm. While the estimate recognizes that typically most companies experienced a much smaller loss (with a one-hour outage costing 56% of all DE, CPM, and FES establishments less than $500), the number was weighted by the much larger losses experienced by a few companies. An estimated 5% of establishments in these sectors incurred costs from a one-hour outage of $20,000 or more, with costs for individual establishments ranging as high as $1.5 million.[20]

EPRI estimated the annual cost of outages across all U.S. business sectors at $104 billion to $164 billion in 2001. With EPRI estimating the average number of power outages in a year over five minutes at 43% (see *Figure 3*), a reasonable estimate for the annual economic cost yields a possible $20 billion to $31 billion (using the Hines estimate of 44% of outages as weather-related) as a lower range. On the higher end (using Mills's estimate of 78% of outages as weather-related), an estimate of the cost yields $35 billion to $55 billion for annual weather-related outage costs.

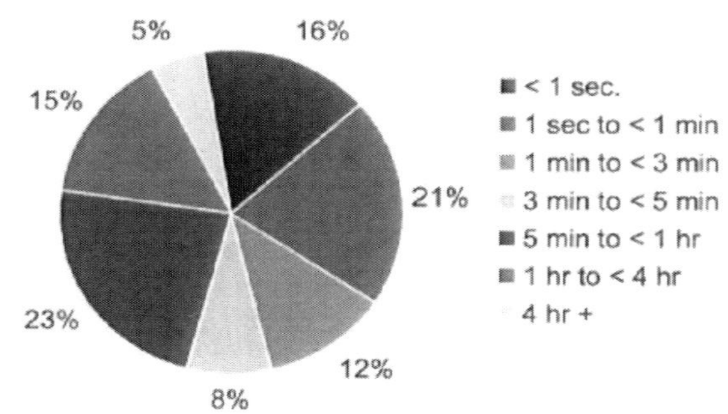

Source: The Cost of Power Disturbances to Industrial & Digital Economy Companies. See http://www.onpower.com/pdf/EPRICostOfPowerProblems.pdf.

Figure 3. Average Length of Power Outages in a Year.

The 2004 LBL report estimated (in its base case) economic costs of sustained power interruptions at $26 billion, with the outages resulting from all causes including weather. In recognition of uncertainties in the data, LBL stated that the true cost of sustained outages could be higher or lower by tens of billions of dollars.[21]

REDUCING STORM-RELATED OUTAGES

Outages are largely a result of damage to distribution systems, which are generally exposed to the elements. Storms can inflict damage to electric power delivery systems in a variety of ways. Wind and rain from a hurricane-type event tends to cause different types of power system delivery failures than snow and ice. This section will look at various scenarios for reducing storm-related outages.

Tree-Trimming Schedules

Many storm-related outages are a result of trees contacting or damaging power delivery systems.

This is a common mode of electric service disruption especially affecting distribution systems (see *Figure 1*). To minimize the chances of power line contacts with trees, utilities typically trim tree branches to maintain a right-of-way[22] free of impinging vegetation.

To maintain reliability of the bulk power system and reduce the risk of outages, NERC has a vegetation management standard which applies to high-voltage transmission lines of 200 kiloVolts (kV) or higher, requiring the transmission line owner to have a vegetation management plan.[23] *Minimum* vegetation clearance distances are calculated by formula, varying from 0.82 feet (for a 69 kV line) to 8.06 feet (for a 765 kV line).[24] NERC's tree-trimming standard does not apply to low-voltage distribution lines, which are typically less than 69 kV.

NERC's reliability rules do not apply to distribution utilities which are commonly subject to state jurisdiction.

Consequently, there is no national law mandating minimum standards for vegetation management around distribution lines.

Schedules for tree-trimming have come under scrutiny as some utilities appear to be increasing the period between these operations as they seek to reduce maintenance costs.

Some state and local jurisdictions have instituted laws or regulations for tree-trimming or vegetation management. For example, New York and several states in New England require electric distribution utilities to submit a plan for tree-trimming and vegetation management for approval by a state utility commission.

> Although each company develops its own plan, subject to state approval, they tend to maintain an eight- to 10-foot clearance around the sides and bottom of a power line and a 12- to 15-foot clearance above it. Trimming cycles range from every four to seven years, with additional trimming and vegetation management for areas that have been problematic in the past.25

Nevertheless, vigilant maintenance of vegetation clearances for power line rights-of-way are generally seen as a good practice that may help to minimize storm-related outages.

Undergrounding of Distribution/Transmission Lines

The practice of putting electricity distribution lines underground has come to be called "undergrounding." In both urban and suburban areas, the practice is common for new construction or in areas being redeveloped to improve aesthetics. Undergrounding is sometimes considered as an option to improve the reliability of service in some areas believed to be susceptible to storm damage.

Underground transmission can be used in urban areas or in other locations where the use of overhead transmission lines may be technically challenging or undesirable. Examples of these situations may include instances when traversing an overland route is difficult, where rivers or natural obstacles exist, or when the line passes through land with a scenic or environmental value judged too great to disturb with an overhead transmission line. Underground transmission may also be less of a regulatory obstacle for these reasons. Underground lines are also safer for wildlife (i.e., a lessened risk of electrical shock) and have far fewer long-term visual impacts.

The cost of burying power lines has been estimated as being from ten to twenty times more expensive than overhead cables,[26] with topography, subsurface conditions, and opportunity to use existing underground conduit among the factors contributing to the differences in cost estimates. While underground lines are less prone to severe weather events, maintenance costs may be higher, especially if the line has to be unearthed or brought to the surface for repair, or if the area is prone to flooding. Thus, the perceived benefits of burying power lines sometimes comes at a cost which communities are not willing to accept.[27] Costs of undergrounding are usually passed along to electricity customers, and collected over an assumed average 25-year service life. Some utilities are also willing to share the cost with communities as they may derive some benefits from undergrounding.[28] However, plans to underground power lines are not always approved by regulatory bodies in some jurisdictions, as electric utilities and transmission builders are often required to consider "least cost" options.

Implementing Smart Grid Improvements

Much of the infrastructure which serves the U.S. power grid is aging. The average age of power plants is now over 30 years, with most of these facilities having a life expectancy of 40 years.[29] Electric transmission and distribution system components are similarly aging, with power transformers averaging over 40 years of age,[30] and 70% of transmission lines being 25 years old or older.[31] As components of the system are retired, they are replaced with newer components often linked to communications or automated systems.

The modernization of the grid to accommodate today's power flows, serve reliability needs, and meet future projected uses is leading to the incorporation of information processing capabilities for power system controls and operations monitoring. The "Smart Grid" is the name given to the evolving electric power network as new information technology systems and capacities are incorporated. A lot has been said about how the Smart Grid can potentially improve electric utility services, but many customers apparently have yet to be convinced that a Smart Grid will benefit them. Smart Grid improvements could help to avoid power outages, as the system would be able to detect problems, and potentially re-route power while alerting system operators to the location of the issue.[32]

Much of the Smart Grid would conceivably be deployed on electricity distribution systems. However, the privacy of customer information has been raised as an issue with smart meters (which relay information related to a customer's electricity usage back to the electric utility). Rate increases have accompanied Smart Grid-related projects in some states, as most of the costs related to smart meters have been directly passed on to electricity customers. Some state jurisdictions are requiring cost-benefit analyses before approving Smart Grid expenditures in rates.[33] Smart Grid programs will likely need to develop good performance metrics for reliability if these new capabilities are to provide the expected service improvements.[34]

Distributed Generation

Distributed generation is located close to the load it is meant to serve. According to DOE, distributed generation "... feeds into the distribution grid, rather than the bulk transmission grid."35 As such, distributed generation includes traditional back-up power sources (such as the large gas-powered generators used by institutions and companies), combined heat and power facilities (used for industrial, district, and community power generation), and renewable electricity power systems used by businesses and residences. Microgrids36 are another form of distributed generation. DOE estimated that as of 2007, there were more than 12 million distributed generation units installed across the United States, with a total capacity of 200 GigaWatts.[37]

Since distributed generation can directly serve the power consumer without external power lines, it is far less vulnerable to the kinds of weather-related service interruptions that can affect utility power systems. DOE stated that “weather is the primary reason for reliability problems, and includes problems caused by lightning strikes, high winds, snowfall, ice, and unexpectedly hot weather. The goal of both planners and operators is to have as resilient a system as possible that can adjust to problems without causing major consequences, and that when outages do occur, they are short-lived and affect the fewest number of customers as possible.”[38] Aside from the potential to provide back-up power to the customer in the event of an outage, DOE found there were other potential benefits to system reliability from distributed generation.[39]

- Several utilities offer financial incentives to owners of emergency power units to make them available to grid operators during times of system need.
- Several regions offer financial incentives or price signals to customers to reduce demand during times of system need (e.g., demand response programs), and some participants in these programs use distributed generation to maintain near-normal on-site operations while they reduce their demand for grid-connected power.

DOE further observed that distributed generation can add diversity to power supply options, thus leading to improvements in overall system adequacy. DOE also stated that distributed generation “has the potential to reduce the number of outages caused by overloaded utility equipment.”[40]

Reliability-Centered Maintenance Regulations

A 2005 study[41] by Davies Consulting for the Edison Electric Institute (EEI) looked at state regulatory investigations of electric utility responses after major storm-related outages.[42] The study came to a number of conclusions concerning utility performance and infrastructure with regard to the time required to restore service after major storms:

- The systems are resilient.
- Adequate funds are being spent on maintenance and reliability.

- Restoration strategies, plans, and practices meet standard utility practices, but more effective resource acquisition (mutual aid) practices need to be employed, such as getting crews on the road earlier and allowing utilities to recover these costs even if they are not used.
- Communication processes with all stakeholders and customers need to be improved—before, during, and after an event.
- Restoration structures and emergency management structures and practices (e.g., Incident Command Structures) need to be aligned.

While the EEI-sponsored study found that adequate amounts were being spent in maintenance and reliability, the study suggested that reliability-centered maintenance[43] (RCM) practices may be an effective tool in preventing some storm-related outages, and could allow for a quicker recovery after a storm-related outage, and encouraged state utility regulators to consider approval of RCM if it was shown to be cost-effective.

> Regulatory efforts, with regard to maintenance, follow two general approaches: development of time-based maintenance intervals with compliance reporting, typically on an annual basis; and development of a higher level maintenance approach with assessments or self-certification. Time-based maintenance may be cost effective for some technologies, but it is likely not effective for all assets. As more utilities evaluate the RCM model, regulators will have to review the benefits and accept the model as part of a cost effective maintenance program. If RCM-type approaches are proven to be effective, but are not accepted by the commissions, utilities will be forced to continue to use the time based methodology and may face a "wall" where assets will need to be replaced in large quantities. As a result, utilities may begin a large-scale and unfocused replacement strategy that does not produce commensurate reliability.[44]

However, the study noted that some state jurisdictions have instituted quality-of-service benchmarks for electric utilities, with penalties imposed on utilities not achieving reliability performance targets. The study observed that penalties rather than rewards would be the focus, as "many state regulators feel that utilities should not be rewarded for service they should already be providing. Second, many regulators believe that there is no tangible benefit to most customers when a utility achieves performance beyond targets."[45] Additionally, those customers willing to pay a premium for reliability (the study reasoned) would likely install their own back-up or distributed generation.

Mutual Assistance Agreements

Depending on the size and extent of a storm causing an outage, a utility may find it necessary to call upon other utilities via its mutual assistance agreements[46] (MAAs) to help it restore services. MAAs can reduce the duration of weather-related outages by bringing in outside resources to aid the recovery effort.

> Mutual assistance is an essential part of the electric power industry's service restoration process and contingency planning. Electric utilities impacted by a major outage event are able to increase the size of their workforce by "borrowing" crews from other utilities. When called upon, a utility will send skilled line workers—both utility employees and contractors—along with specialized equipment to help with the restoration efforts of a fellow utility.
>
> ... Mutual Assistance includes primary and secondary response with back-up contingency plans. Contingency plans are developed by the smaller, more regionalized RMAGs (Regional Mutual Assistance Groups). These plans assure that resource transfers have a minimal effect on a regional area in case an unexpected event occurs. Contingency plans can be done in various ways depending on the amount of resources being transferred either from a region or a larger geographic area such as across North America.[47]

MAAs are voluntary arrangements made with other electric utilities in states or regions unlikely to be affected by the same storm or weather-related event. As such, it may take time for MAAs to result in additional help arriving, as the utility workers must travel sometimes significant distances. Assuming that roadways are open, these convoys must travel at speeds which can be safely navigated by the bucket trucks and other specialized vehicles making the journey.

EEI provides advice on how to structure and arrange MAAs to its member utilities. While MAAs are voluntary and not regulated, they have come under state regulatory scrutiny especially in longer-duration outages. But the extent of damage can slow the response of other utilities to respond to calls for assistance, as a utility's first obligation is to restore service to its own customers in the event of an outage. Nevertheless, effective MAAs are invaluable to the recovery effort after a major storm-related outage.

CONCLUDING OBSERVATIONS

In any discussion of storm-related power outages, two prominent themes emerge—preparation and recovery. If utilities are aware of an impending storm or weather-related event which may cause outages, they are expected to make preparations for restoration of services in as timely a manner as possible. Recovery from any such event will depend on the severity of the storm and the resulting damage. However, recovery can be hastened, and the amount of damage to electric power infrastructure can be minimized, if good maintenance, restoration, organization, and communications strategies are followed on an ongoing basis. Electric utilities should also be aware of local weather-related events in the past. If an area is prone to particular seasonal extremes of weather, then utilities would be expected to work with state and local regulatory authorities to look at ways to reduce vulnerabilities of the power and related infrastructure.

The cost of many aged parts of the electric power system have already been recovered in rates by electric utilities. The time for new investments may be approaching (if it is not already here) considering the projected future uses of the grid. Power delivery systems are most vulnerable to storms and extreme weather events. Improving the overall condition and efficiency of the power delivery system can only serve to improve the resiliency of the system, and help hasten recovery from weather-related outages. Ultimately, however, electric utilities are responsible for this infrastructure. They are in the business of selling electricity, and they cannot sell electricity if their power delivery systems are out of service.

Options for Congress

The inconsistency of data from outage reporting is an issue in quantifying the impacts of storm-related and other power outages. Congress could empower the National Institute of Standards or DOE or some other federal agency to develop standards for the consistent reporting of power outage data.

While responsibility for the reliability of the bulk electric system is under FERC, no central responsibility exists for distribution systems. One possible option would be to bring these systems under the ERO for reliability purposes. However, this would require changes in law, with the Federal Power Act[48] being key to any changes.

Formalizing review or even coordinating MAAs under a federal agency like the Federal Emergency Management Agency or the Department of Homeland Security may be an option that Congress may want to consider. This would not necessarily mean federal approval of MAAs, but may help in the cooperative coordination of additional federal and state resources, especially in a wide, multi-state weather event. The utility industry, however, may be opposed to government involvement in MAAs.

The aged condition of the electrical grid in many parts of the United States will likely only detract from and compound recovery efforts. While there has been much discussion of transmission system inadequacies and inefficiencies, many distribution systems are in dire need of upgrades or repairs. The cost of upgrading the grid to meet future uses is expected to be high, as the estimates of the ASCE show, and the importance of a robust grid to U.S. economic performance goes without question. While the federal government recently made funding available for specific Smart Grid projects or transmission lines under the American Recovery and Reinvestment Act of 2009,[49] there has not been a comprehensive effort to study the needs, set goals, and provide targeted federal funding for the modernization of the U.S. grid as part of a long-term national energy strategy. Such an effort would also require decisions about the appropriate roles of government and the private sector.

End Notes

[1] A power outage results in a complete loss of power for the end-use customer.

[2] P.L. 109-58.

[3] 16 U.S.C. § 824o.

[4] Alison Silverstein, Transmission 101, National Association of Regulatory Utility Commissioners, April 20, 2011, p. 30, http://www.naruc.org/grants/Documents/Silverstein%20NCEP%20T-101%20042011 1.pdf.

[5] Paul Hines, Jay Apt, and Sarosh Talukdar, Trends in the History of Large Blackouts in the United States, University of Vermont, 2008, http://www.uvm.edu/~phines/publications/2008/Hines_2008_blackouts.pdf.

[6] Kristina Hamachi LaCommare and Joseph H. Eto, Understanding the Cost of Power Interruptions to U.S. Electricity Consumers, Ernest Orlando Lawrence Berkeley National Laboratory, September 2004, http://certs.lbl.gov/pdf/ 55718.pdf.

[7] Per the definition of a "sustained interruption" used by the Institute of Electrical and Electronics Engineers (IEEE). Guide for Electric Power Distribution Reliability Indices, No. 1366.

[8] "... widespread power losses resulting from major natural events (primarily storms but also hurricanes and earthquakes) are sometimes not included in the same data categories as more routine power losses. As a result, power losses from natural events are not always included in data used for cost estimates." LaCommare and Eto, op. cit., p. 5.

[9] U.S. Department of Energy, Energy Information Administration, Form OE-417. See http://www.eia.gov/cneaf/ electricity/page/disturb_events.html.

[10] Evan Mills, Extreme Grid Disruptions and Extreme Weather, Lawrence Berkeley National Laboratory, U.S. DisasterReanalysis Workshop, May 3, 2012, http://evanmills.lbl.gov/presentations/Mills-Grid-Disruptions-NCDC3May2012.pdf.

[11] E-mail from Evan Mills, Lawrence Berkeley National Laboratory, August 30, 2012.

[12] Council of European Energy Regulators, 4th Benchmarking Report on Quality of Electricity Supply 2008, C08-EQS24-04, December 10, 2008, http://www.energy-regulators.eu/portal/page/portal/EER_HOME/EER_PUBLICATIONS/CEER_PAPERS/Electricity/2008/C08-EQS-24-04_4th%20Benchmarking%20Report%20EQS_10-Dec-2008_re.pdf.

[13] Ibid.

[14] Ibid, p. 9.

[15] Ibid, page 7.

[16] "The cumulative need, based on anticipated investment levels and the estimated investment gap, will be $673 billion by 2020, an average of about $75 billion per year. Based on investment over the past decade, closing the gap is within reach: the average annual need projected from 2012 through 2020 falls within the range of annual investment totals in the last decade, and there is not a single year through 2020 that is projected to be outside that range." American Society of Civil Engineers, Failure to Act: The Economic Impact of Current Investment Trends in Electricity Infrastructure, 2011, http://www.asce.org/uploadedFiles/Infrastructure/Failure_to_Act/energy_report_FINAL2.pdf.

[17] "Based on current investment trends, the national electricity infrastructure gap is estimated to be $107B by 2020, or just over $11B per year. By 2020, shortfalls in grid investments are expected to account for almost 90% of the investment gap with nearly $95B in additional dollars needed to modernize the grid." Ibid.

[18] Consortium for Electric Infrastructure for a Digital Society, The Cost of Power Disturbances to Industrial & Digital Economy Companies, June 29, 2001, http://www.onpower.com/pdf/EPRICostOfPowerProblems.pdf.

[19] "The market value of goods and services produced by labor and property in the United States, regardless of nationality." See http://www.bea.gov/glossary/glossary_g.htm.

[20] Consortium for Electric Infrastructure for a Digital Society, op. cit., p. 16.

[21] LaCommare and Eto, op. cit., Executive Summary.

[22] A strip of land occupied by active electricity transmission or distribution facilities.

[23] North American Electric Reliability Corporation, Transmission Vegetation Management NERC Standard FAC-003-2 Technical Reference, September 2009, http://www.nerc.com/docs/standards/sar/FAC-003- 2_White_Paper_2009Sept9.pdf.

[24] Clearances at sea level. Ibid, page 47.

[25] Lee R. Hansen, Utility Tree Trimming in Other States, State of Connecticut, December 19, 2011, http://www.cga.ct.gov/2011/rpt/2011-R-0459.htm.

[26] Jeff Griffin, "Underground Electric Transmission Installations Gaining Traction," Underground Construction, vol. 65, no. 6, June 2010, http://www.underground constructionmagazine.com/underground-electric-transmissioninstallations-gaining-traction.

[27] Julie Patel, "Are You Willing to Pay Thousands to Keep the Power on After a Storm?," Sun Sentinel, June 25, 2011, http://articles.sun-sentinel.com/2011-06-25/business/fl-fpl-underground-lines-20110624_1_power-lines-utility-linestamara-tennant.

[28] Ken Silverstein, "Upgrading the Grid by Going Underground," Forbes, August 2, 2012, http://www.forbes.com/sites/ kensilverstein/2012/08/02/upgrading-the-grid-by-going-under ground/2/.

[29] Massachusetts Institute of Technology, Retrofitting of Coal-Fired Power Plants for CO2 Emissions Reductions, March 23, 2009, http://web.mit.edu/mitei/docs/reports/meeting-report.pdf.

[30] Thomas A. Prevost and David J. Woodcock, Transformer Fleet Health and Risk Assessment, Weidman Electrical Technology, IEEE PES Transformers Committee Tutorial, March 13, 2007, http://grouper.ieee.org/groups/ transformers/info/S07/S07-TR_LifeExtension.pdf.

[31] K. Anderson, D. Furey, and K. Omar, Frayed Wires: U.S. Transmission System Shows Its Age, Fitch Ratings, October 25, 2006.

[32] Ken Silverstein, op. cit.

[33] Katherine Tweed, Illinois Rejects Ameren's Smart Grid Plan, Greentech Media, May 30, 2012, http://www.greentechmedia.com/articles/read/illinois-rejects-amerens-smart-grid-plans1/.

[34] Galvin Electricity Initiative, Electricity Reliability: Problems, Progress and Policy Solutions, February 2011, http://www.galvinpower.org/sites/default/files/Electricity_ Reliability_ 031611.pdf.

[35] U.S. Department of Energy, The Potential Benefits of Distributed Generation and Rate-Related Issues That May Impede Their Expansion, February 2007, http://www.ferc. gov/legal/fed-sta/exp-study.pdf.

[36] A microgrid is any small or local electric power system that is independent of the bulk electric power network. For example, it can be a combined heat and power system based on a natural gas combustion turbine (which cogenerates electricity, and hot water or steam from water used to cool the natural gas turbine), or diesel generators, renewable energy, or fuel cells. A microgrid can be used to serve the electricity needs of data centers, colleges, hospitals, factories, military bases, or entire communities (i.e., "village power").

[37] "Over 99% of these units are small emergency reciprocating engine generators or photovoltaic systems, installed with inverters that do not feed electricity directly into the distribution grid. However, ... this large number of smaller machines represents a relatively small fraction of the total installed capacity." U.S. Department of Energy, The Potential Benefits of Distributed Generation and Rate-Related Issues That May Impede Their Expansion, February 2007, http://www.ferc.gov/legal/fed-sta/exp-study.pdf, p. 33.

[38] U.S. Department of Energy, The Potential Benefits of Distributed Generation and Rate-Related Issues That May Impede Their Expansion, February 2007, http://www.ferc.gov /legal/fed-sta/exp-study.pdf, p. 42.

[39] Ibid.

[40] Ibid. p. 50.

[41] Davies Consulting, Inc., State of Reliability Distribution Regulation in the United States, Edison Electric Institute, September 2005, http://legalectric.org/f/2010/04/ stateof distributionreliability-2005.pdf.

[42] The events included the North Carolina and South Carolina ice storm of December 2003, Hurricanes Isabel and Juan (September 2003), the Utah snow storm of December 2004, four hurricanes in Florida in summer 2004, and the Maritime ice storm of November 2004.

[43] "Reliability centered maintenance ... is a scientific process designed to develop a maintenance program, focuses on equipment condition and is matched to the wear-out pattern of that

equipment based on operation, not time. Those industries using reliability centered maintenance such as the U.S. airline industry, the U.S. Navy (including nuclear submarines), nuclear power plants, and fossil fuel plants, indicate that adoption of RCM has increased reliability while simultaneously reducing costs." Davies Consulting, Inc., op. cit., p. 21.

[44] Ibid.

[45] Ibid.

[46] Most utilities have found that they cannot staff up internally to provide the manpower they need to respond to major storms. Therefore, they rely on mutual assistance agreements where they agree to share line crews and equipment to help each other respond to major storms. See http://www.eei.org/ourissues/electricitydistribution/Documents/ StormRestore Report.pdf.

[47] See http://www.eei.org/ourissues/ElectricityTransmission/ Reliability/Pages/ElectricSector MutualAssistance.aspx.

[48] 16 U.S.C. § 824o.

[49] The American Recovery and Reinvestment Act of 2009 (P.L. 111-5) made available $3.4 billion for funding Smart Grid projects, and approximately $6.5 billion in borrowing authority for transmission systems was authorized for the Western Area Power Administration and the Bonneville Power Administration. Additionally, $6 billion in loan guarantees was made available for renewable energy and electric power transmission.

INDEX

D

E

F

S

T

U

V

W

Y